Religion *in* POLITICS

Religion *in* POLITICS

Constitutional and Moral Perspectives

MICHAEL J. PERRY

New York Oxford • Oxford University Press • 1997

Oxford University Press

Oxford New York
Athens Auckland Bangkok Bogota Bombay Buenos Aires
Calcutta Cape Town Dar es Salaam Delhi Florence Hong Kong
Istanbul Karachi Kuala Lumpur Madras Madrid Melbourne
Mexico City Nairobi Paris Singapore Taipei Tokyo Toronto

and associated companies in
Berlin Ibadan

Published by Oxford University Press, Inc.
198 Madison Avenue, New York, NY 10016

Oxford is a registered trademark of Oxford University Press

Library of Congress Cataloging-in-Publication Data
Perry, Michael J.
Religion in politics : constitutional and moral perspectives /
Michael J. Perry.
p. cm.
Includes index.
ISBN 0-19-510675-X
1. Religion and Politics—United States. 2. Religion and state—United States. 3. United States—Religion. 4. United States—Constitutional law—Moral and ethical aspects. 5. Law and ethics.
I. Title.
BL65.P7P47 1997
322'.1'0973—dc20 96-23812

9 8 7 6 5 4 3 2 1

Printed in the United States of America
on acid-free paper

For my mother,
Mary Frances Gregory Perry

CONTENTS

Religion *in* POLITICS

Introduction

RELIGION IN POLITICS

If few Americans were religious believers, the issue of the proper role of religion in politics would probably be marginal to American politics, because religion would be marginal to American politics. But most Americans are religious believers. Indeed, the citizenry of the United States is one of the most religious—perhaps even the most religious—citizenries of the world's advanced industrial democracies. According to recent polling data, "[a]n overwhelming 95% of Americans profess belief in God";[1] moreover, "70% of American adults [are] members of a church or synagogue."[2] If there were, among the vast majority of Americans who are religious believers, a consensus about most religious matters, the issue of the proper role of religion in politics would probably engage far fewer Americans than it does, because few Americans would have to fear being subjected to alien religious tenets. But there is, among American believers, a dissensus about many fundamental religious matters, including many fundamental religious-moral matters. Because the United States is *both* such a religious country *and* such a religiously pluralistic country (now more than ever), the issue of the proper role of religion in politics is anything but marginal to American politics. The proper role of religion in politics is a central, recurring issue in the politics of the United States.

In this book, I address a fundamental question about religion in politics: What role may religious arguments play, if any, either in public debate about what political choices to make or as a basis of political choice? Two phrases appear throughout my discussion: "religious arguments" and "political choices". Political choices are not all of the same kind. The political choices with which I am principally concerned in this book are those that ban or otherwise disfavor one or another sort of human conduct based on the view that the conduct is immoral. A law banning abortion is a paradigmatic instance of the kind of political choice I have in mind; a law banning homosexual sexual conduct is another. The religious arguments with which I am principally concerned here are arguments that one or another sort of human conduct, like abortion or homosexual sexual conduct, is immoral. By a "religious" argument, I mean an argument that relies on (among other things) a religious belief: an argument that presupposes the truth of a religious belief and includes that belief as one of its essential premises. The belief that God exists—"God" in the sense of a transcendent reality that is the source, the ground, and the end of everything else—is a "religious" belief, as is a belief about the nature, the activity, or the will of God.

The controversy about the proper role of religious arguments in politics comprises two debates: a debate about the *constitutionally* proper role of religious arguments in politics and a related, but distinct, debate about their *morally* proper role.[3] According to the constitutional law of the United States, government may not "establish" religion. Given this "nonestablishment" norm, what role is it constitutionally proper (permissible) for religion to play, if any, in the politics of the United States? In particular:

- Does a legislator or other public official, or even an ordinary citizen, violate the nonestablishment norm by presenting a religious argument in public debate about what political choice to make? For example, does a legislator violate the nonestablishment norm by presenting, in public debate about whether the law should recognize homosexual marriage, a religious argument that homosexual sexual conduct is immoral?
- Does a political choice violate the nonestablishment norm if it is made on the basis of a religious argument? For example, does a law banning abortion violate the nonestablishment norm if it is based even partly on a religious argument that abortion is immoral?

Beyond the constitutional inquiry lies the moral inquiry. That an act does not violate any constitutional norm does not mean that the act is morally appropriate. Constitutional legality does not entail, much less equal, moral propriety.

- Even if, as I believe to be the case, neither citizens nor even legislators or other public officials violate the nonestablishment norm by presenting religious arguments in public political debate (i.e., in public debate about what political choices to make), this question remains: All things considered, is it morally appropriate for citizens—in particular, legislators and other public officials—to present such arguments in public political debate?

Moreover, constitutional illegality does not entail, much less equal, moral impropriety. That an act would violate a constitutional norm does not entail that the act would be, apart from its unconstitutionality, morally inappropriate.

- Even if, as I believe to be the case, a political choice would violate the nonestablishment norm if no plausible secular argument supported it, this question remains: Apart from the nonestablishment norm, is it morally appropriate for citizens and legislators and other public officials to rely on religious arguments in making a political choice even if, in their view, no persuasive or even plausible secular argument supports the choice?[4]

I first addressed such questions in *Love and Power: The Role of Religion and Morality in American Politics*.[5] In the years since *Love and Power* was published, and partly in response to critical commentary on *Love and Power*, I have continued to think about the difficult problem of religion in politics. As it happens, my thinking has been a rethinking. Among the inquiries I pursue in this book are two that I neglected to pursue in *Love and Power*: first, the question of the constitutionality, under the nonestablishment norm, of religious arguments in politics; second, the question of the relationship between religiously based moral arguments and secular moral arguments. Partly in consequence of pursuing those two important inquiries, I have been led to revise substantially the position on religion in politics I presented in *Love and Power*. It bears mention that nothing in this book presupposes that the reader is familiar with *Love and Power*.

An overview of the three chapters of this book might be useful. I begin, in chapter 1, by discussing the freedom of religion protected by the constitutional law of the United States. My principal aims in chapter 1 are two. First, I present what seems to me to be the most sensible account of the general meaning of the two basic constitutional norms regarding religion: the free exercise norm and, especially, the nonestablishment norm. (Along the way, I address such controversial constitutional issues as prayer in public schools, government display of religious symbols, and government aid to religiously affiliated education.) Then, drawing on my account of the meaning of the nonestablishment norm, I make an argument about the role it is constitutionally permissible for religious arguments—in particular, religious arguments about the morality of human conduct—to play in the politics of the United States. I conclude that neither citizens nor even legislators or other public officials violate the nonestablishment norm by presenting religious arguments in public political debate, but that a political choice would violate the norm if no plausible secular argument supported it. Again, my principal concern in this book are political choices about the morality of human conduct.

Then, in chapters 2 and 3, I turn from the constitutional inquiry about religion in politics to the moral inquiry. Because it is focused on the constitutional law of the United States, chapter 1 is relevant to questions about "religion in politics" as they arise in the United States. Chapters 2 and 3, however, are relevant to such questions as they arise in any democratic political community that, like the United States, is religiously pluralistic. My response to the moral inquiry, in chapters 2 and 3, does not depend on my response, in chapter 1, to the constitutional inquiry. One might disagree with some or all of what I have to say in chapter 1 and yet agree with some or all of what I have to say in chapters 2 and 3.

In chapter 2, I conclude that as a matter of political morality it is not merely permissible but important that religious arguments be presented—principally, so that they can be tested—in public political debate. Two of the most important contributors to the debate about the morally proper role of religion in politics are Kent Greenawalt and John Rawls.[6] Greenawalt and Rawls have each defended a position less congenial to the airing of religious arguments in public political debate than the position I defend in this book. I explain, in chapter 2, why I disagree both with Greenawalt's position and with Rawls's.

In chapter 3, I distinguish between the two basic sorts of religious argument about the morality of human conduct: religious argument about human worth and religious argument about human well-being. I then argue that in making a political choice about the morality of human conduct, citizens and legislators and other public officials may rely on a religious argument that all human beings, and not merely some (e.g., white persons), are sacred *even if*, in their view, no persuasive secular argument supports the claim that all human beings are sacred. (As Ronald Dworkin has noted, although the term "sacred" is often used in a theistic sense, it is not invariably used in that sense. "Sacred" may also be used, and sometimes is used, in a nonreligious—secular—sense.[7]) Next, I argue that in making a political choice about the morality of human conduct, citizens and legislators and other public officials should not rely on a religious argument about the requirements of human well-being unless, in their view, a persuasive secular argument reaches the same conclusion about those requirements as the religious argument.

The political controversy in the United States today about the morality of homosexual sexual conduct—which is at the center of the debate about whether the law should recognize homosexual marriage or at least grant some sort of marriage-like status to same-sex unions—is a principal context for the debate about the proper role of religion in politics. In chapter 3, I use that controversy to illustrate the central argument of the chapter: that in making a political choice about the morality of human conduct, legislators and others should not rely on a religious argument unless, in their view, a persuasive secular argument reaches the same conclusion about the requirements of human well-being. (As I explain in chapter 3, an argument, whether religious or secular, that all homosexual sexual conduct is immoral is at bottom an argument about the requirements of human well-being.) Is there a persuasive secular argument that all homosexual sexual conduct is immoral? John Finnis recently tried to construct a secular argument in support of the traditional religious tenet that all homosexual sexual conduct—even homosexual sexual conduct that is embedded in and expressive of a lifelong, monogamous relationship of faithful love—is immoral, but, as I demonstrate in chapter 3, Finnis's secular argument is not sound. In the wake of Finnis's failure, one can fairly doubt that any secular argument that all homosexual sexual conduct is immoral is sound. If one agrees that no such secular argument is sound, then, for the reasons I develop in chapter 3, one should not rely on a religious argument that all such conduct is immoral as a basis of political choice, least of all a coercive political choice.

In discussing drafts of this book with various persons and groups, I have occasionally been asked about the "voice" that informs my conception of the proper role of religion in politics. I have written this book as a Christian.[8] In particular, I have written it as a Catholic Christian thoroughly imbued with the spirit of the Second Vatican Council (1962–65).[9] But I have written this book as a Christian who is extremely wary of the God-talk in which most Christians (and many others) too often and too easily engage; I have written it, that is, in the spirit of apophatic Christianity.[10] Moreover, I have written this book as one who stands between all religious nonbelievers on the one side and many religious believers—especially theologically conservative believers—on the other. Religious nonbelievers, many of whom would like to marginalize the role of religious discourse in public political debate, are the principal addressees of my argument, in chapter 2, that it is not merely permissible but important that religious arguments about the morality of human conduct be presented in public political debate. Religious nonbelievers are also the principal addressees of my argument, in chapter 3, that in making a political choice about the morality of human conduct, legislators and others may rely on a religious argument that all human beings are sacred even if, in their view, no persuasive secular argument supports the claim about the sacredness of all human beings. By contrast, religious believers—especially Christians, who are, by far, still the largest group of religious believers in the United States—are the principal addressees of my argument, in chapter 3, that in making a political choice about the morality of human conduct, especially a coercive political choice, legislators and others should not rely on a religious argument about the requirements of human well-being unless, in their view, a persuasive secular argument reaches the same conclusion about those requirements as the religious argument.

The comments with which I conclude this book, at the end of chapter 3, are meant mainly for theologically conservative Christians, many of whom are likely to be skeptical about the part of my argument, in chapter 3, addressed principally to religious believers. (It is precisely in that part of my argument, which concerns mainly claims about what God has revealed about the requirements of human well-being, that my wariness about God-talk is most engaged.) Nonetheless, I hope that my concluding comments will also speak, if only indirectly, to theologically conservative members of other religious traditions as well.[11]

Acknowledgments

For helpful comments or discussion as I was writing this book, I am grateful to many friends and colleagues, especially Robert Audi, Thomas Berg, William Collinge, Daniel Conkle, Charles Curran, Franklin Gamwell, Kent Greenawalt, Stephen Gardbaum, Scott Idleman, Andrew Koppelman, William Kralovec, Douglas Laycock, Daniel Morrissey, Mark Noll, John Rawls, Richard Saphire, Rodney Smith, David Smolin, Lawrence Solum, Laura Underkuffler, Howard Vogel, Gerry Whyte, and Ashley Woodiwiss. I am also grateful to have had the opportunity to discuss a draft of this book in several venues during the 1995–96 academic year: Wheaton College (Illinois); the Northwestern University Center for the Humanities; the Saint Thomas University School of Law; the Cumberland School of Law of Samford University; the University of Colorado School of Law; the University of California at Davis School of Law; the University of San Diego School of Law; and the St. John's University School of Law. Finally, I am grateful to the Northwestern University law students who, in 1995 and 1996, joined me in thinking about "Religion, Politics, and the Constitution".

Chapter 1

THE CONSTITUTIONAL LAW OF RELIGIOUS FREEDOM

In this chapter, I present what seems to me to be the most sensible account of the general meaning of the two basic constitutional norms regarding religion: the free exercise norm and the nonestablishment norm. Then, drawing on that discussion, I make an argument about the role it is constitutionally permissible for religious arguments—in particular, religious arguments about the morality of human conduct—to play in the politics of the United States.[1]

The heart of my account of the free exercise and nonestablishment norms is that government may not make judgments about the value or disvalue—the truth value, the moral value, the social value, any kind of value—of religions or religious practices or religious (theological) tenets. The proposition that government should not have the power to make such judgments is the yield of a long history, leading right up to the present, that has taught us that when the politically powerful arrogate to themselves the power to make judgments about the value or disvalue of religions or religious practices or tenets, they usually do harm, sometimes much harm, and little if any good—and that, moreover, it is utterly unnecessary, in terms of achieving the objectives proper to government, which we can subsume under the term "the common good", that the politically powerful have such power. The power to make judgments about the value or disvalue of religion is, in the hands of government—especially in a religiously pluralistic context like our own—an understandably tempting but nonetheless unnecessary and ultimately dangerous power. This lesson of history, which can be affirmed at least as vigorously by those of us who count ourselves religious believers as by those who do

not, lies at the very foundation of the American commitment to the free exercise and nonestablishment norms, which together constitute the particular conception of religious freedom that is the subject of this chapter.

I. Getting from There to Here

According to the First Amendment to the Constitution of the United States, "Congress shall make no law respecting an establishment of religion, or prohibiting the free exercise thereof; or abridging the freedom of speech, or of the press; or the right of the people peaceably to assemble, and to petition the government for a redress of grievances." Thus, the First Amendment makes two statements concerning religion: that "Congress shall make no law respecting an establishment of religion" and that "Congress shall make no law . . . prohibiting the free exercise [of religion]".[2] Yet, the freedom of religion protected by the constitutional caselaw of the United States—the constitutional law developed by the Supreme Court of the United States in the course of resolving conflicts about what the Constitution forbids—is much broader than the language of the First Amendment indicates. It is not just Congress that may not make a law respecting an establishment of religion or prohibiting the free exercise thereof; no branch of the national government—legislative, executive, or judicial—may take action establishing a religion or prohibiting the free exercise thereof. Moreover, it is not just the national government that may not take such action; no state government may do so either. According to the constitutional law of the United States, neither any branch of the national government nor any state government may either establish religion or prohibit the free exercise thereof.[3]

Is there a basis in any part of the Constitution for extending the command of the First Amendment beyond Congress to the other two branches of the national government? Of course, the First Amendment concerns not just religion, but also speech, press, and peaceable assembly. Still, the command of the First Amendment is directed only at *Congress*—and even then only at congressional *legislation*. "'Congress' does not on its face refer to the president, the courts, or the legions that manage the Executive Branch, and 'law' only arguably includes administrative orders or congressional investigations. Freedom of political association, which (without serious controversy) has been held to be fully protected, is not even mentioned in the document . . . . It requires a theory to get us where the courts have gone."[4] Is there such a "theory"?[5] The Ninth Amendment states: "The enumeration, in the Constitution, of certain rights, shall not be construed to deny or disparage others retained by the people." If the Constitution protects a right to a fundamental human freedom, like the

freedom of religion or the freedom of speech, against one branch of the national government (e.g., Congress), perhaps a right to that freedom against the other two branches ought to be deemed an unenumerated constitutional right "retained by the people"—*unless* there is a reason for concluding that the freedom, even though it is already constitutionally protected against the one branch of the national government, ought not to be constitutionally protected against the other two branches. However, this "theory" for getting us a part of the way from the language of the First Amendment to the constitutional law of the United States regarding religion, speech, press, and peaceable assembly will not persuade everyone, because it relies on a controversial claim about the meaning of the Ninth Amendment. (I have discussed the controversy about the meaning of the Ninth Amendment elsewhere.[6])

Is there a basis in any part of the Constitution for extending the command of the First Amendment beyond Congress and the rest of the national government to the governments of the states? The second sentence of section one of the Fourteenth Amendment states: "No State shall make or enforce any law which shall abridge the privileges or immunities of citizens of the United States; nor shall any State deprive any person of life, liberty, or property, without due process of law; nor deny to any person within its jurisdiction the equal protection of the laws." It has been argued that this language—in particular, the privileges or immunities clause—was meant to protect against the states (among the other things it was meant to protect against the states) the "privileges" and "immunities" already protected against the national government by the Bill of Rights (and, moreover, to protect them against state government in the same way and to the same extent they were already protected against the national government). Recent arguments to that effect are powerful, in my view.[7] Nonetheless, not everyone is persuaded.[8] More important, that section one of the Fourteenth Amendment was meant to protect against state government the privileges and immunities already protected against the national government by the Bill of Rights does not entail that the First Amendment was meant to protect any privileges or immunities against the national government. Jay Bybee has recently mounted an important argument that the First Amendment, unlike the Second through Eighth Amendments, was not meant to protect any privileges or immunities but only to make explicit a congressional disability—a lack of legislative power on the part of Congress.[9] The point is not that there is no support for the position the Supreme Court has embraced for most of the twentieth century, namely, that section one of the Fourteenth Amendment was meant to "incorporate" the First Amendment's ban on government establishing religion, prohibiting the free exercise thereof, or abridging the freedom of speech or the freedom of the press, thereby making it applicable to the states.[10] The point, rather, is that whatever support there is, is not unequivocal.

One might want to inquire, therefore, whether, *if* there is no basis in the text of the Constitution for extending the command of the First Amendment beyond the national government to state government, it is *nonetheless* possible to justify the Supreme Court's insistence that neither the national government nor state government may establish religion, prohibit the free exercise thereof, or abridge the freedom of speech or the freedom of the press. (Similarly, one might want to inquire whether, if there is no textual basis for extending the command of the First Amendment beyond Congress to the rest of the national government, it is nonetheless possible to justify the Court doing so.) I have pursued such an inquiry elsewhere.[11] For present purposes, however, it suffices to observe that this aspect of constitutional caselaw has come to be widely and deeply affirmed by "We the people of the United States".[12] Indeed, the proposition that neither the national government nor state government constitutionally may establish religion, prohibit the free exercise thereof, or abridge the freedom of speech or the freedom of the press has come to be a virtual axiom of American political-moral culture. The proposition is so deeply embedded in the American way of life that, as a practical matter, it is irreversible. Moreover, and as I have explained elsewhere, the proposition is, for "We the people" now living, constitutional bedrock.[13] No one who contended that the Supreme Court should overrule the proposition would be taken seriously as a nominee to the Court.

The freedom of religion protected by the constitutional law of the United States, then, is the freedom constituted by two norms: first, that government shall not establish religion (which I will call "the nonestablishment norm") and, second, that government shall not prohibit the free exercise of religion ("the free exercise norm"). There are different conceptions of (and, correspondingly, different institutionalizations of) "freedom of religion". The conception to which the United States is committed, by its fundamental law, is the conception constituted by the free exercise and nonestablishment norms. Although, as I said, little controversy attends the question whether the nonestablishment and free exercise norms should be maintained as constitutional norms[14]—that is, as *judicially enforceable* constitutional norms—there is some controversy about precisely what the norms forbid government to do.

II. Free Exercise

What does it mean to say that government may not prohibit the free exercise of religion? The "exercise" of religion comprises many different but related kinds of religious practice, including: public affirmation of religious beliefs; affiliation, based on shared religious beliefs, with a church or other religious group; worship and study animated by religious beliefs; the proselytizing dissemination of religious beliefs or other religious information; and moral choices, or even a whole way of life, guided by religious beliefs.[15] It is implausible to construe the free exercise norm to forbid government to prohibit each and every imaginable religious practice—including, for example, human sacrifice. Indeed, by its very terms the norm forbids government to prohibit, not the exercise of religion, but the "free" exercise of religion. Just as goverment may not abridge "the freedom of speech" or "the freedom of the press", so too it may not prohibit the "freedom" of religious exercise.[16] It is one thing to argue, as I do in section V of this chapter, that the freedom of religious exercise includes more than just freedom from discrimination against religious practice. It is another thing altogether, and insane, to suggest that the freedom of religious exercise is an unconditional right to do, on the basis of religious belief, whatever one wants. One need not concoct frightening hypotheticals about human sacrifice to dramatize the point. One need only point, for example, to the refusal of Christian Science parents to seek readily available lifesaving medical care for their gravely ill child.[17] Just as the freedom of speech is not a license to say whatever one wants wherever one wants whenever one wants and the freedom of the press is not a license to publish whatever one wants wherever one wants whenever one wants, so, too, the freedom of religious exercise is not a license to do, on the basis of religious belief, whatever one wants wherever one wants whenever one wants. If, then, there are some religious practices that the free exercise norm does not forbid government to prohibit, what does the norm forbid government to do?

Some persons might fear, dislike, or otherwise disvalue a religious practice. According to the free exercise norm, however, government may not "prohibit"—it may not ban or otherwise impose any regulatory restraint on—any religious practice *as such*, that is, any practice embedded in and expressive of one or more religious beliefs.[18] More generally, government may not take any action, impeding any religious practice, based on the view that the practice is, as religious practice, inferior along one or another dimension of value to another religious or nonreligious practice or to no practice at all. Thus, the free exercise norm is an antidiscrimination provision: It forbids government to take prohibitory action discriminating against religious practice (i.e., disfavoring religious practice as such). Government "would be 'prohibiting the free exercise [of religion]' if it sought to ban [acts or refusals to act] only when they are engaged in for religious reasons, or only because of the religious belief that they display. It would doubtless be unconstitutional, for example, to ban the casting of 'statues that are to be used for worship purposes,' or to prohibit bowing down before a golden calf."[19]

To ban or otherwise impose a regulatory restraint on a religious practice (as such) is one way for government to take prohibitory action discriminating against the practice, but it is not the only way. The free exercise norm, construed in an appropriately generous way, forbids indirect as well as direct restraint. Government may not impede a religious practice (as such) by denying to persons, because they engage in the practice, a benefit it would otherwise confer on them. It simply would make no sense, it would be naive and even foolish, to forbid government to ban "the casting of 'statues that are to be used for worship purposes'" while leaving government free to deny to persons, because they cast such statues, a benefit it would otherwise give to them.

That the free exercise norm is an antidiscrimination provision—that government "prohibits" the free exercise of religion if its prohibitory action disfavors religious practice as such—is not controversial, but that the norm is more than an antidiscrimination provision is controversial. There is disagreement, both on the Supreme Court and off, about whether the norm forbids government to do anything other than take prohibitory action discriminating against religious practice. Does the free exercise norm forbid, in addition to prohibitory action that is discriminatory, at least some governmental action that, although *nondiscriminatory*, impedes religious practice? Or is it problematic to read the language of the free exercise norm—in particular, the word "prohibit"—to extend even to some nondiscriminatory governmental action impeding religious practice? In order to pursue this inquiry, we should first answer a prior question: What does the nonestablishment norm forbid government to do?

III. Nonestablishment

What does it mean to say that government may not establish religion; at least, what has it come to mean, in the United States? As I stated at the beginning of this chapter, the central point of the free exercise and nonestablishment norms, taken together, is that government may not make judgments about the value or disvalue—the truth value, the moral value, the social value—of religions or religious practices or religious (theological) tenets as such (i.e., as religious). Government has no such power, and government may not arrogate to itself any such power. Whereas the free exercise norm forbids government to take prohibitory action disfavoring one or more religious practices as such, the nonestablishment norm forbids government to favor one or more religions as such; in particular, it forbids government to discriminate in favor of membership in one or more churches or other religious communities or in favor of the practices or tenets of one or more churches. No matter how much some persons might prefer one or more religions, government may not take any action based on the view that the preferred religion or religions are, as religion, better along one or another dimension of value than one or more other religions or than no religion at all. So, for example, government may not take any action based on the view that Christianity, or Roman Catholicism, or the Fifth Street Baptist Church, is, as a religion or a church, closer to the truth than one or more other religions or churches or than no religion at all—or, if not necessarily closer to the truth, at least a more authentic reflection of the religious history and culture of the American people. (One might believe that by reflecting the religious history and culture of a people, laws and other political choices achieve a moral authority they would otherwise lack and help to maintain a useful sense of moral solidarity among the people.) Similarly, no matter how much some persons might prefer one or more religious practices, government may not take any action based on the view that the preferred practice or practices are, as religious practice (practice embedded in and expressive of one or more religious beliefs), better—truer or more efficacious spiritually, for example, or more authentically American—than one or more other religious or nonreligious practices or than no religious practice at all. For example, government may not take any action based on the view that the Lord's Prayer is, as prayer, better than one or more other prayers or than no prayer at all. Finally, no matter how much some persons might prefer one or more religious tenets—that is, one or more tenets about the existence, nature, activity, or will of God—government may not take any action based on the view that the preferred tenet or tenets are truer or more authentically American or otherwise better than one or more competing religious or nonreligious tenets. For example, government may not take any action based on the view that the Roman Catholic doctrine of apostolic succession (which is, whatever else it is, a doctrine that presupposes that God exists and makes a claim about God's activity and/or will) is closer to the truth than one or more competing theological or nontheological doctrines. (My reference in the preceding two sentences is to religious tenets other than religiously based tenets about the morality of human conduct. I discuss the implications of the nonestablishment norm for religiously based moral tenets in section VII of this chapter.)

Like the free exercise norm, then, the nonestablishment norm is an antidiscrimination provision; but unlike the free exercise norm, which forbids government to engage in prohibitory action that *disfavors* one or more religious practices as such, the nonestablishment norm forbids government to *favor* one or more religions as such—including one or more religious practices or tenets.[20] Government may not, for example, subsidize the casting of "statues that are to be used for worship purposes". Moreover, the nonestablishment norm, like the free exercise norm, forbids indirect as well as direct action. Government may not give to persons, because they engage in a particular religious practice, a benefit it would otherwise deny to them, if doing so is based on the view that the practice is, as religious practice, better than one or more other religious or nonreligious practices or than no religious practice at all. It would make no sense to forbid government to subsidize the casting of "statues that are to be used for worship purposes" while leaving government free to confer special benefits on persons because they cast such statues.

IV. Why Nonestablishment?

Why is it a good thing that the constitutional law of the United States includes the free exercise norm? Because, as history helps us understand, and as the protection of religious freedom by the international law of human rights suggests, it would be a bad thing, indeed a terrible thing, were government in the United States free to discriminate against one or more religions or religion generally. Acts of governmental discrimination against religion, in the United States and elsewhere, have been and almost surely would continue to be illiberal, if not authoritarian or totalitarian, in their motivation and divisive and sometimes even destabilizing in their consequences. Moreover, a politics free to discriminate against religion is a politics free to secure favors from one or more religions, including the favor of genuflecting before the politically powerful, in return for not discriminating against them. What an ugly state of affairs that would be: some religions discriminated against and some other religions bought off. Finally, there is simply no need for government in the United States to be free to discriminate against religion; nothing of real value is lost by forbidding government to discriminate against religion.

But is it a good thing that the constitutional law of the United States includes not just the free exercise norm, but the nonestablishment norm as well? Note, in that regard, that although the international law of human rights includes provisions that function just like the free exercise norm—in particular, forbidding government to discriminate against religious practice—the international law of human rights does not include anything like a nonestablishment norm. Nor does governmental action that would violate the nonestablishment norm necessarily violate any human right.

Consider, for example, the Constitution of Ireland, which, in the Preamble, affirms a nonsectarian Christianity: "In the name of the Most Holy Trinity, from Whom is all authority and to Whom, as our final end, all actions both of men and States must be referred, we, the people of Éire, humbly acknowledging all our obligations to our Divine Lord, Jesus Christ, Who sustained our fathers through centuries of trial, . . . do hereby adopt, enact, and give to ourselves this Constitution."[21] Given the religious commitments of the vast majority of the people of Ireland, it is not at all surprising that the Irish Constitution affirms Christianity. In so doing, the Irish Constitution violates no human right. Three things are significant here. First, the religious convictions implicit in the Irish Constitution's affirmation of Christianity in no way deny—indeed, they affirm—the idea that *every* human being, *Christian or not*, is sacred; they affirm, that is, the very foundation of the idea of human rights.[22] Second, the Irish Constitution's affirmation of Christianity is not meant to insult or demean anyone; it is meant only to express the most fundamental convictions of the vast majority of the people of Ireland.[23] Third, and perhaps most importantly, the Irish Constitution protects the right, which is a human right, to freedom of religion; moreover, it protects this right not just for Christians, who are the vast majority in Ireland, but for all citizens. Article 44 states, in relevant part: "Freedom of conscience and the free profession and practice of religion are . . . guaranteed to every citizen. . . . The State shall not impose any disabilities or make any discrimination on the ground of religious profession, belief or status."[24] Therefore, the conclusion that in affirming Christianity the Irish Constitution violates a human right—or that in consequence of the affirmation Ireland falls short of being a full-fledged liberal democracy—is, in a word, extreme.[25] Something Brian Barry said recently, in *Justice as Impartiality*, is relevant here:

> We must, of course, keep a sense of proportion. The advantages of establishment enjoyed by the Church of England or by the Lutheran Church in Sweden are scarcely on a scale to lead anyone to feel seriously discriminated against. In contrast, denying the vote to Roman Catholics or requiring subscription to the Church of England as a condition of entry to Oxford or Cambridge did constitute a serious source of grievance. Strict adherence to justice as impartiality would, no doubt, be incompatible with the existence of an established church at all. But departures from it are venial so long as nobody is put at a significant disadvantage, either by having barriers put in the way of worshipping according to the tenets of his faith or by having his rights and opportunities in other matters (politics, education, occupation, for example) materially limited on the basis of his religious beliefs.[26]

The question arises, therefore, whether, even though it is certainly a good thing—a very good thing—that the constitutional law of the United States includes the free exercise norm, it is also a good thing—and, if so, how good—that it includes the nonestablishment norm.[27] Let's come at the question from the other side: Why might one think it a bad thing that the constitutional law of the United States includes the nonestablishment norm? What need is there for government in the United States to be free to discriminate in favor of one or more religions? What of value is lost, if anything, by forbidding government to discriminate in favor of religion?

It is at least a moderately good thing, in my view, that the constitutional law of the United States includes the nonestablishment norm, because, all things considered, it would be at least a moderately bad thing—and perhaps, from time to time, a very bad thing—were either the national government or state government constitutionally free to discriminate in favor of one or more religions or religious practices or tenets.[28] Were the politically powerful free to discriminate in favor of one or more religions, history suggests that they would almost certainly do little if any good and some, or more than some, harm. That they would do little if any good and at least some harm is especially likely, of course, in a society as religiously pluralistic as the United States.

It bears emphasis, at this point, that one important way to protect freedom of religion is to protect, as the nonestablishment norm does, freedom *from* governmentally imposed religion—which is, of course, a freedom of religious believers no less than of religious nonbelievers. An important way to protect the freedom of those of us who count ourselves religious to follow our religious consciences where they lead—especially the freedom of those of us who are not politically powerful—is for the constitutional law of the United States to forbid the politically powerful among us to act, in large ways or small, in obvious ways or subtle, to privilege ("establish") their brand of religion. Thus, the nonestablishment norm is good news not just for the atheists and agnostics among us; it is good news for us all. It is noteworthy, too, that the nonestablishment norm protects not only freedom of religion, but also religion itself. One way for government to corrupt religion—to co-opt it, to drain it of its prophetic potential—is to seduce religion to get in bed with government; an important way to protect religion, therefore, is to forbid government to get in bed with religion. "In this framework religion becomes a tool, a means to control behavior, an instrument to revivify the people, a cheap hireling to provide a basis for unity, a means merely to achieve political ends. In the end, religion is the loser. True religion, genuine faith, is defamed, desecrated, and trivialized. This is the lesson of history, yet we are on the verge of repeating the same error. Religious belief has its public dimensions, to be sure, but it is first and foremost a matter of private right. Church-state separation is the great protector of true faith, not its inhibitor."[29]

Moreover—and this is perhaps the crux of the matter—there is simply *no practical need* for political bureaucracies, even democratic political bureaucracies, to have the power to discriminate in favor of religion. As Doug Laycock has observed: "In the case of religion, no one has to rule. There is no need for the government to make decisions about Christian rituals versus Jewish rituals versus no religious rituals at all. For government to make that choice is simply a gratuitous statement about the kind of people we really are. By making such statements, the government says the real American religion is watered-down Christianity, and everybody else is a little bit un-American."[30] From a practical standpoint, it is utterly unnecessary to leave the politically powerful free to discriminate in favor of one or more religions or religious practices or tenets—it is utterly unncessary, that is, in terms of achieving the objectives proper to government, which we may subsume under the term "the common good". Because, as our historical experience attests, the politically powerful would almost certainly do little if any good and at least some harm if they had that power to discriminate, and because it is gratuitous to leave them with it, they should not have such power.

Would it be a good thing, or at least not such a bad thing, were government free at least to discriminate, not in favor of one or more religions in relation to one or more other religions, but in favor of religion generally? As I explain in section VI, for the constitutional law of the United States to adopt the accommodation position is for it to discriminate in favor of religion—specifically, in favor of religious practice—generally. Except for the good arguably achieved by adopting the accommodation position, I am skeptical that there is anything of real value we Americans might reasonably hope to accomplish by discriminating, or by leaving our politicians free to discriminate, in favor of religion generally. Why do we need to enact laws or make political choices that discriminate in favor of religion generally? From a practical standpoint at least, there is no more need to discriminate in favor of religion generally than to discriminate in favor of one or more religions in relation to one or more other religions. Let us remember, at this juncture, that "a governmental policy of remaining neutral in regard to religion is [not] a denial of religious truth; it is only a commitment to steer clear of a dimension of life about which government has little competence, leaving the pursuit of religious truth to every individual. We would do well to heed the words of James Madison: 'The religion of every man must be left to the conviction and confidence of every man. In matters of religion no man's right is to be abridged by the institution of civil society; religion is wholly exempt from its competence.' These words . . . embody the true meaning of 'religious equality.'"[31]

V. Nonestablishment Conflicts

My aim in section II and III was to give, briefly and somewhat abstractly, an account of what the free exercise and nonestablishment norms, correctly understood, forbid government to do. Of course, what seems to me to be the correct understanding of the norms might not be what seems to the Supreme Court of the United States to be the correct understanding. As it happens, however, what I have said about the free exercise norm—that it is, whatever else it is, an antidiscrimination provision, forbidding government to take prohibitory action disfavoring religious practice as such—accurately represents the position not only of a majority of the members of the Court today, but of the whole Court.[32] (Again, there is disagreement both on the Court and off about whether the free exercise norm is more than an antidiscrimination provision. More about that disagreement in the next section.) Moreover, what I have said about the nonestablishment norm substantially represents the position of the Court today.[33] Consider, in particular, the Court's stance with respect to four of the most politically charged nonestablishment conflicts to have engaged its attention in the last fifty years.

Prayer in Public Schools

For an agency of government, including the public schools, to institutionalize prayer, whether a particular prayer or a particular kind of prayer, is almost certainly for the agency to take action based on the view that the prayer or kind of prayer is, as prayer, better—truer or more efficacious spiritually, for example, or more authentically American—than another prayer or kind of prayer or than no prayer at all. Worse, for the public schools to take such action is "coercive" in the sense and to the extent that, as a practical matter, some students or teachers or both must be present at a religious service—prayer—they would prefer to avoid.[34] Although it would be very problematic (and controversial) to make "coercion" an essential element of a nonestablishment violation,[35] that governmental action is coercive means that the action is especially offensive in terms of the ideal of nonestablishment. If and to the extent the action is coercive, government is saying, in effect: "It is *so* important that there be prayer, here and now—prayer, here and now, is *so* much better than the absence of prayer—that there shall be prayer, here and now, even though that means that some of you must be present at a religious service you would prefer to avoid."[36] The Supreme Court agrees that it violates the nonestablishment norm for the public schools to institutionalize or otherwise orchestrate prayer—and it agrees that to the extent such governmental action is coercive, the action is especially offensive.[37]

Government Display of Religious Symbols

For government to display a religious symbol[38] is not necessarily for government to discriminate in favor of the religion whose symbol it is. Imagine, for example, that during December a municipality displays a crèche and a menorah in the town square. Is such action based on the view that Christianity and Judaism are, as religions, better (e.g., truer, more authentically American, etc.) than one or more other religions or than no religion at all? Not necessarily. We need to know more, we need to take into account all the relevant particularities of context. Are the crèche and the menorah the only symbols displayed? Or are some secular symbols of the season displayed alongside the religious symbols—a figure of Santa Clause, for example, or a Christmas tree, or both? Government display of a religious symbol or symbols might indeed violate the nonestablishment norm,[39] but it does so if, and only if, the display is based on the view that one or more religions (or religious practices or tenets) are, as such, better than one or more other religions or than no religion at all. Of course, there might be room for a reasonable difference in judgments about whether such a display is so based. It seems doubtful, however, that a municipality would not have displayed, in the town square during the winter holiday season, a crèche and a menorah alongside a figure of Santa Clause and a Christmas tree *but for* the view that Christianity and Judaism are, as religions, better along one or another dimension of value than one or more other religions or than no religion at all. It seems more likely that such a scene would be displayed in the town square even absent the view that Christianity and Judaism are, as religions, better; it seems likely that such a scene would be displayed on the basis of the view that *all* the major symbols of the season—a season that for some is religious, for others, secular, and for still others, both religious and secular—should brighten the town square and contribute, each symbol in its own way, to the festive, generous spirit of the holidays. This is not to claim that the view that Christianity and Judaism are, as religions, better is necessarily absent, but only that such a view is probably not a basis of the display—or, at least, that it would be very difficult to conclude with confidence that it *is* a basis of the display. In any event, the position I have sketched here with respect to government display of religious symbols is substantially the position of the Supreme Court.[40]

Evolution and Creation Science

Recall that, according to the construal of the nonestablishment norm I have presented, government may not take action based on the view that that one or more religious tenets are closer to the truth or otherwise better than one or more competing religious or nonreligious tenets. An Arkansas statute, enacted in 1928, made "it unlawful for a teacher in any state-supported school or university 'to teach the theory or doctrine that mankind ascended or descended from a lower order of animals,' or 'to adopt or use in any such institution a textbook that teaches' this theory."[41] Did the statute violate the nonestablishment norm? Yes, because the inference is irresistible—especially given the history of the statute, which "was a product of the upsurge of 'fundamentalist' religious fervor of the twenties"[42]—that the statute was based on the view that a particular religious tenet (about the material origins of the human species) is, as religious doctrine, truer or otherwise better than the scientific theory whose teaching the statute makes unlawful. Here "better" means something like "more reflective of, and therefore less subversive of, the dominant religious beliefs of the community". As the Court said in *Epperson v. Arkansas*: "[T]here can be no doubt that Arkansas has sought to prevent its teachers from discussing the theory of evolution because it is contrary to the belief of some that the Book of Genesis must be the exclusive source of doctrine as to the origin of man. No suggestion has been made that Arkansas' law may be justified by considerations of state policy other than the religious views of some of its citizens. It is clear that fundamentalist sectarian conviction was and is the law's reason for existence."[43]

A Louisiana statute prohibited "the teaching of the theory of evolution in public schools unless accompanied by instruction in 'creation science.' No school is required to teach evolution or creation science. If either is taught, however, the other must also be taught. The theories of evolution and creation science are statutorily defined as 'the scientific evidences for [creation and evolution] and inferences from those scientific evidences.'"[44] Did the statute violate the nonstablishment norm? It certainly did if, as a majority of the Court concluded in *Edwards v. Aguilard*, the Louisiana statute was based on a view like that on which the Arkansas statute struck down in *Epperson* was based. The serious question, as Justice Scalia (joined by Chief Justice Rehnquist) argued in a dissenting opinion, was whether the Louisiana statute was indeed based on such a view or, instead, on the view that both of the competing scientific theories of creation should be taught "so that students would not be 'indoctrinated' but would instead be free to decide for themselves, based upon a fair presentation of the scientific evidence, about the origin of life."[45]

Government Aid to Religiously Affiliated Schools

Does it violate the nonestablishment norm for a state to give financial aid (e.g., in the form of vouchers) to religiously affiliated schools—in particular, to religiously affiliated elementary and secondary schools?[46] It depends. Is the aid given only to nonpublic schools that are religiously affiliated, or is it given to all nonpublic schools (that meet the state's criteria for schools of the sort in question)? Assume that although the aid is given to all nonpublic schools, the vast majority of such schools are Roman Catholic. Does the legislative history or other probative material suggest that although the aid is given to all nonpublic schools, the aid program was established based on the view—that it would not have been established but for the view—that Roman Catholicism is, as religion, better than one or more other religions or than no religion at all? We can well imagine that a state, like my own state of Illinois, with a city, like Chicago, might well be persuaded to provide some minimal amount of financial aid to all nonpublic elementary and secondary schools—or even just to all nonpublic elementary and secondary schools that are, first, financially ailing and, second, serving children from poor neighborhoods. A state might provide such aid based partly on the view that Catholic elementary and secondary schools are an invaluable educational resource for the community and cannot be permitted to fail for lack of adequate financial support—which is a quite plausible view, especially in cities like Chicago. But it seems farfetched to suppose that if today a state would provide such aid, it would invariably do so based on the view that Roman Catholicism (or Christianity generally) is, as religion, better than other religions or than no religion. Yet, only the latter view, not the former, is constitutionally problematic under the nonestablishment norm.

There is a virtual consensus among commentators that the Supreme Court's decisions about government aid to religiously affiliated schools—in particular, the Court's chaotic collection of decisions about aid to religiously affiliated elementary and secondary schools—are an unholy mess. It is difficult to discern a nonarbitrary distinction between the situations in which the Court has invalidated government aid to religiously affiliated elementary or secondary schools and the situations in which it has not.[47] Happily, there now seem to be five members of the Court—enough for a majority—willing to simplify the Court's jurisprudence in this area, and to do so in a way consistent with what I am suggesting here, namely, that government may give financial aid to religiously affiliated schools—including elementary and secondary schools—if, and only if, first, the criteria for such aid are religiously neutral and, second, the aid program is not a subterfuge for affirming one or more religions as such.[48]

To hold that government may not give financial aid to religiously affiliated schools even if the criteria for such aid are religiously neutral and the aid program is not a subterfuge for affirming one or more religions as such would be to hold that government must discriminate against religious activities. That government may not discriminate *in favor of* religious activities does not entail that government must discriminate *against* them. I cannot imagine why, as a matter of political morality, one would want to require government to discriminate against religious activities—or, therefore, why one would want to construe the nonestablishment norm to require such discrimination. (Indeed, I cannot imagine why one would want to permit government to discriminate against religious activities.) The fact that some persons object to their tax monies being spent to aid religious activities no more justifies such discrimination—it no more justifies according constitutional status to their objection, however conscientious their objection might be—than the fact that some persons object to their tax monies being spent to aid military activities, or abortion, justifies according constitutional status to *their* objection. If anything, an objection to one's tax monies being spent to aid an activity that one believes, and can reasonably believe, to be immoral—for example, abortion—ought to be taken *more* seriously, not less, than an objection to one's tax monies being spent to aid an activity that one happens not to endorse but that, nonetheless, one cannot reasonably believe to be immoral—for example, a church sponsoring an elementary school. Of course, because government may not discriminate in favor of religious activities under the nonestablishment norm, it may not spend tax monies to aid religious activities as such—that is, as *religious* activities. But, again, that government may not discriminate in favor of religious activities, that it may not favor religious activities as such, does not mean that government must, or even should, discriminate against religious activities; it does not mean that government must, or should, disfavor religious activities as such.[49] That it is wrong for government to discriminate in favor of an activity does not mean that it is right for government to discriminate against the activity, any more than that it is wrong for government to discriminate against an activity means that it is right for government to discriminate in favor of the activity. Given the extent to which the citizenry of the United States is religious,[50] it seems perverse to suggest that the nonestablishment norm should be construed not only to forbid government to discriminate in favor of religious activities, but also to require government to discriminate against such activities.[51] Moreover, such discrimination—and, therefore, such a construal of the nonestablishment norm—seems plainly inconsistent with the free exercise norm, which, whatever else it is, is an antidiscrimination provision that forbids government to disfavor religious activities as such. As the Court put it in 1947, the free exercise and nonestablishment norms together require "the state to be neutral in its relationship with groups of religious believers and non-believers; it does not require the state to be their adversary. State power is no more to be used to handicap religions than it is to favor them."[52]

VI. Free Exercise, Nonestablishment, and the Problem of "Accommodation"

To rehearse an earlier point: That the free exercise norm is an antidiscrimination provision—that government "prohibits" the free exercise of religion if its prohibitory action disfavors religious practice as such—is not controversial, but that the norm is more than an antidiscrimination provision *is* controversial. As I said, there is disagreement, both on the Supreme Court and off, about whether the norm forbids government to do anything other than take prohibitory action discriminating against religious practice. Does the free exercise norm forbid, in addition to prohibitory action that is *discriminatory*, at least some governmental action that, although *nondiscriminatory*, impedes religious practice? One position—the "accommodation" position—holds that the free exercise norm not only forbids government to discriminate against religious practice, but also requires government to maximize the space for religious practice by exempting religious practice from an otherwise applicable ban or other regulatory restraint that would interfere substantially with a person's ability to engage in the practice, unless the exemption would seriously compromise an important public interest. So, for example, if a state were to ban the consumption of alcoholic beverages, the state would have to exempt from the ban the consumption of wine in the Christian sacrament of the Eucharist unless the exemption would seriously compromise an important public interest. In that sense, government must sometimes "accommodate" religious practice, by exempting it from an otherwise applicable regulatory restraint, even if the failure to exempt would not constitute discrimination against the practice. According to the accommodation position, government "prohibits" the free exercise of religion not only if its prohibitory action disfavors religious practice as such, but also if it fails to exempt religious practice from a (nondiscriminatory) restraint if the exemption would not seriously compromise an important public interest.

Some constitutional scholars contend for the accommodation position, others contend against it.[53] As the accommodation debate illustrates, the conventional range of reference of the word "prohibit" is broad enough to accommodate either the accommodation position or the rejection of the position. In that sense, the language of the free exercise norm underdetermines the choice between the accommodation position and its rejection. By government shall not "prohibit" the free exercise of religion one might mean only that government shall not take prohibitory action that discriminates against religious practice, that disfavors religious practice as such. Or, instead, one might mean *both* that government shall not take prohibitory action discriminating against religious practice *and* that it shall not take action that, though nondiscriminatory, substantially interferes with a person's ability to engage in a religious practice if exempting the practice would not seriously compromise an important public interest. However, that the conventional range of reference of the word "prohibit" underdetermines the choice between the accommodation position and its rejection does not mean that the free exercise norm that has been established as a part of the constitutional law of the United States is not the norm for which the accommodation position argues.

What norm was the free exercise language of the First Amendment ("Congress shall make no law . . . prohibiting the free exercise [of religion]") understood to communicate by the "We the people of the United States" over two hundred years ago who, through their representatives, put that language of the First Amendment into the text of the Constitution; that is, what directive did they understand themselves to be issuing —what norm did they mean to establish?[54] Although it is not unequivocal, the historical evidence does seem to support an answer congenial to the accommodation position.[55] As I explained at the beginning of this chapter, however, the First Amendment norm that "Congress shall make no law . . . prohibiting the free exercise [of religion]" is not the same as the free exercise norm, now a bedrock part of the constitutional law of the United States, that neither any branch of the national government nor any state government shall prohibit the free exercise of religion. Who are the persons who made the free exercise norm a part of the constitutional law of the United States? Even if it is not clear who they are—even if it is controversial who they are—it is quite clear that they are not the same as the "We the people of the United States" over two hundred years ago who, through their representatives, made the First Amendment, including its free exercise language, a part of the Constitution of the United States. If we accept the "incorporation" position sketched earlier in this chapter, and if we accept that the norms incorporated and thereby made applicable to the states include the First Amendment's ban on government establishing religion, prohibiting the free exercise thereof, or abridging the freedom of speech or the freedom of the press, then the persons who made the free exercise norm a part of the constitutional law of the United States—the broad free exercise norm that is a limitation on state government as well as on the national government—are the "We the people" in the 1860s who, through their representatives, made the Fourteenth Amendment a part of the Constitution. According to Kurt Lash, the particular free exercise norm that (in Lash's scenario) they made a part of the Constitution supports the accommodation position.[56] If we credit Lash's argument, that norm is to this effect: Government shall not take prohibitory action discriminating against religious practice or action that, though nondiscriminatory, substantially interferes with a person's ability to engage in a religious practice if exempting the practice would not seriously compromise an important public interest.

Even if, contra Lash and others, the Fourteenth Amendment was not meant to make the First Amendment applicable to the states,[57] it is, as I said, constitutional bedrock for "We the people of the United States" now living that the First Amendment's ban on government establishing religion, prohibiting the free exercise thereof, or abridging the freedom or speech or the freedom of the press applies to the states. It is not constitutional bedrock, however, that the free exercise norm applicable to the states (and to the national government) supports the accommodation position. (For those for whom historical arguments about "original meaning" or "original understanding" have great authority in constitutional adjudication,[58] it must be noted that such arguments for the accommodation position are controversial.[59]) And, as it happens, a majority of the Supreme Court, in 1990, rejected the accommodation position. In *Employment Division, Oregon Department of Human Resources v. Smith*, the Court wrote that "if prohibiting the exercise of religion . . . is . . . merely the incidental effect of a generally applicable and otherwise valid provision," the free exercise norm, without regard to whether the refusal to exempt the religious practice in question serves an important public interest, has not been violated.[60] The five-person majority's rejection (in an opinion by Justice Scalia) of the accommodation position provoked not only the other four persons on the Court,[61] but also so many interested persons off the Court that an unprecedented alliance of groups, from the American Civil Liberties Union on the one side to the so-called "religious right" on the other, joined forces to lobby Congress to undo the decision. The alliance was successful. Congress enacted, in 1993, the Religious Freedom Restoration Act (RFRA),[62] the declared purpose of which is to reaffirm the accommodationist version of the free exercise norm and, in particular, to make that version a part of the statutory law of the United States. According to section 3 of RFRA: "Government shall not substantially burden a person's exercise of religion even if the burden results from a rule of general applicability, [unless] . . . it demonstrates that application of the burden to the person (1) is in furtherance of a compelling governmental interest; and (2) is the least restrictive means of furthering that compelling governmental interest."[63]

The fundamental political-moral rationale for the accommodation position is that the "free exercise of religion" is such an important value that government must not only not discriminate against religious practice but must also do what it can, short of compromising an important public interest, to avoid putting substantial impediments in the way of religious practice. A similar but broader rationale, not limited to religious practice, explains the breadth of the freedom of conscience protected by the international law of human rights: freedom not only from discrimination but also from some nondiscriminatory interference. The International Covenant on Civil and Political Rights,[64] which the United States ratified in 1992, provides, in Article 18, that: "Everyone shall have the right to freedom of thought, conscience, and religion. This right shall include freedom to have or to adopt a religion or belief of his choice, and freedom, either individually or in community with others and in public or private, to manifest his religion or belief in worship, observance, practice and teaching." Article 18 then states: "Freedom to manifest one's religion or beliefs may be subject only to such limitations as are prescribed by law and are necessary to protect public safety, order, health, or morals or the fundamental rights and freedoms of others."[65] Virtually identical provisions inhabit the Universal Declaration of Human Rights (Articles 18 and 29),[66] the Declaration on the Elimination of All Forms of Intolerance and of Discrimination Based on Religion or on Belief (Article 1),[67] the American Convention on Human Rights (Article 12),[68] and the European Convention on Human Rights (Article 9).[69] According to each of these instruments, government must not only not discriminate against religious or other conscientious practice but must also avoid interfering with such practice except to the extent "necessary to protect public safety, order, health, or morals or the fundamental rights and freedoms of others." (The Declaration on the Elimination of All Forms of Intolerance and of Discrimination Based on Religion or on Belief, which I have reproduced in the appendix to this chapter, is the most elaborate articulation of the human right to freedom of religion in international law.)

Earlier I said that whether it is problematic to construe the word "prohibit" to extend even to some nondiscriminatory governmental action impeding religious practice depends in part on what the nonestablishment norm forbids government to do. Recall that, among other things, the nonestablishment norm forbids government to take action based on the view that one or more religious practices are, as religious practice, better than one or more other religious or nonreligious practices or than no religious practice at all. The principal argument against the accommodation position is that for government to do what the accommodation position requires—favor, by accommodating, religious practice *as such*—is for government, in violation of the nonestablishment norm, to take action based on the view that, at least as a general matter, religious practices are, as such, better or more valuable than nonreligious practices.[70] Now, it might be ideal if the constitutional law of the United States were revised to protect acts of secular conscience on a par with acts of religious conscience.[71] (A utopian suggestion? As we've seen, the international law of human rights *already* protects not only acts of religious conscience, but acts of conscience generally.) As it stands, however, the free exercise norm protects, from discrimination, only acts of religious conscience. It is one thing to suggest that the constitutional law of the United States be revised to protect acts of conscience generally. It is another thing altogether, and extreme, to suggest that unless and until the constitutional law of the United States is revised to protect acts of conscience generally, the free exercise norm should be ignored.[72] The free exercise norm, no less than the nonestablishment norm, is a bedrock part of the existing constitutional law of the United States, and as such it operates as a practical limit on what we may reasonably construe the nonestablishment norm to forbid. At the same time, the nonestablishment norm, in conjunction with the international law of human rights, might exert a gravitational pull on our understanding of what the constitutional law of the United States *should* protect. Eventually, we might come to accept that the constitutional law of the United States should, to some extent, include protection not only for acts of religious conscience, but also for acts of conscience generally—and for the processes of conscience formation.[73]

Meanwhile, however, it makes sense to conclude that it is not inconsistent with the nonestablishment norm for constitutional law to require government to accommodate only acts of religious conscience (with the proviso, of course, that government need not accommodate even an act of religious conscience if to do so would seriously compromise an important public interest). After all, because the free exercise norm (as a bedrock part of the existing constitutional law of the United States) operates as a practical limit on what we can reasonably construe the nonestablishment norm to forbid, it is not inconsistent with the nonestablishment norm for constitutional law to protect only acts of religious conscience from discrimination against them. Is it less extreme to insist that accommodating only acts of religious conscience violates the nonestablishment norm than to insist that protecting only acts of religious conscience from discrimination against them violates the nonestablishment norm? It is important to keep things in perspective here: According to the accommodationist version of the free exercise norm, no act of religious conscience merits exemption from a truly nondiscriminatory ban or other regulatory restraint *if the exemption would compromise an important public interest.*[74]

If the accommodationist version of the free exercise norm were morally obnoxious, even one who found the historical argument for the accommodationist version of the norm stronger than the historical argument for the anti-accommodationist version might want to find a respectable way to reject the accommodationist version. But although the broader position represented by the international law of human rights—the position that protects not just acts of religious conscience, but acts of conscience generally—might be morally preferable, it seems to me farfetched to claim that the accommodation position is morally obnoxious. Indeed, one might fairly conclude that in the context of the United States, the accommodationist version of the free exercise norm is morally preferable to the anti-accommodationist version, according to which the free exercise norm is only an antidiscrimination provision. By insisting that the free exercise norm protects religious practice only from prohibitory action disfavoring religious practice as such—by insisting that the norm is only an antidiscrimination provision or, as the Court later put it, a "nonpersecution principle"[75]—the Court in *Employment Division, Department of Human Resources of Oregon v. Smith*[76] "started down a doctrinal path that leads to a constitutional discourse in which contending parties accuse each other of hostility, persecution, and bad faith. . . . [T]his sort of demonizing debate is precisely what a doctrinal emphasis on motive as a dispositive factor is calculated to elicit. . . . [I]f one is searching for alternatives, then pre-*Smith* free exercise jurisprudence—not merely the "compelling interest" balancing test . . . but rather the discourse of humility and tolerance exemplified in [*Wisconsin v. Yoder*]—invites renewed consideration."[77]

I said earlier that according to the nonestablishment norm, government may not confer on persons, because they engage in a particular religious practice, a benefit it would otherwise deny to them, if doing so is based on the view that the practice is, as religious practice, better than one or more other religious or nonreligious practices or than no religious practice at all. Accepting the accommodationist version of the free exercise norm back into the constitutional law of the United States, as we should, would require adding this proviso: "*unless* (a) the benefit is in the form of an exemption from an otherwise applicable ban or other regulatory restraint that would substantially burden the practice and (b) the exemption does not seriously compromise an important public interest."[78]

VII. Religion in Politics: Constitutional Perspectives

I have articulated, in this chapter, the basic features of the freedom of religion, including the freedom of religious believers and nonbelievers alike from governmentally imposed religion, protected by the constitutional law of the United States—the freedom of religion constituted by the free exercise and nonestablishment norms. The stage is now set for us to pursue the inquiry about the role it is constitutionally permissible for religious arguments to play, if any, in the politics of the United States. As I noted in the preface to this book, the controversy about the proper role of religious arguments in politics comprises two inquiries: an inquiry about the *constitutionally* proper role of religious arguments in politics and a related but distinct inquiry about their *morally* proper role. In the next two chapters, I pursue the moral inquiry. In the remainder of this chapter, I address this question: Given the freedom of religion protected by the constitutional law of the United States—given, in particular, the nonestablishment norm—what role, if any, is it constitutionally permissible for religious arguments to play, in the United States, either in public debate about what political choices to make or as a basis of political choice?

As a prelude both to the remainder of this chapter and to the next two chapters, let me rehearse some clarifications I made in the preface.

- The political choices with which I am principally concerned in this book are those that ban or otherwise disfavor one or another sort of human conduct based on the view that it is immoral for human beings (whether all human beings or some human beings) to engage in the conduct. A law banning abortion is a paradigmatic instance of the kind of political choice I have in mind; a law banning homosexual sexual conduct is another.
- The religious arguments with which I am principally concerned are arguments that one or another sort of human conduct, like abortion or homosexual sexual conduct, is immoral.
- By a "religious" argument, I mean an argument that relies on (among other things) a religious belief: an argument that presupposes the truth of a religious belief and includes that belief as one of its essential premises. A "religious" belief is, for present purposes, either the belief that God exists—"God" in the sense of a transcendent reality that is the source, the ground, and the end of everything else—or a belief about the nature, the activity, or the will of God.[79] A belief can be "nonreligious", then, in one of two senses. The belief that God does not exist is nonreligious in the sense of "atheistic". A belief that is about something other than God's existence or nonexistence, nature, activity, or will is nonreligious in the sense of "secular". In addition to religious arguments, we can imagine both "atheistic" arguments and "secular" arguments. One who is "agnos-

> tic" about the existence of God—who neither believes nor disbelieves that God exists—will find only secular arguments persuasive.[80]

Let's begin with this question: Does a legislator or other public official,[81] or even an ordinary citizen, violate the nonestablishment norm by presenting a religious argument in public political debate? For example, does a legislator violate the nonestablishment norm by presenting, in public debate about whether the law should recognize homosexual marriage, a religious argument that homosexual sexual conduct is immoral? An affirmative answer is wildly implausible. Every citizen, without regard to whether she is a legislator or other public official,[82] is constitutionally free to present in public political debate whatever arguments about morality, including whatever religious arguments, she wants to present. Indeed, the freedom of speech protected by the constitutional law of the United States is so generous that it extends even to arguments, including secular arguments, that may not, as a constitutional matter, serve as a basis of political choice—for example, the argument that persons of nonwhite ancestry are not truly or fully human (which is an unconstitutional basis of political choice under the antidiscrimination part of the Fourteenth Amendment[83]). Thus, whether or not religious arguments may, as a constitutional matter, serve as a basis of political choice, it is clear that citizens and even legislators and other public officials are constitutionally free to present such arguments in public political debate.[84] The nonestablishment norm is not to the contrary.[85]

Moreover, to disfavor religious arguments relative to secular ones would violate the core meaning—the antidiscrimination meaning—of the free exercise norm. After all, included among the religious practices protected by the free exercise norm are bearing public witness to one's religious beliefs and trying to influence political decisionmaking on the basis of those beliefs.[86] As the Second Vatican Council of the Catholic Church observed in the document *Dignitatis Humanae*, true freedom of religion includes the freedom of persons and groups "to show the special value of their doctrine in what concerns the organization of society and the inspiration of the whole of human activity."[87] Although the nonestablishment norm, as I have explained, forbids any branch or agency of government to do certain sorts of things, to engage in certain sorts of actions, it does not forbid any person—including any person who happens to be a legislator or other public official—to say whatever she wants to say, religious or not, in public political debate. The serious question, then, is not whether legislators or other public officials, much less citi-

zens, violate the nonestablishment norm by presenting religious arguments in public political debate.[88] The serious question, rather, is whether government would violate the nonestablishment norm by basing a political choice—for example, a law banning abortion—on a religious argument.[89]

Recall that among the other things it forbids government to do, the nonestablishment norm forbids government to take any action based on the view that one or more religious tenets are closer to the truth or more authentically American or otherwise better than one or more competing religious or nonreligious tenets. For example, government may not base any action on the view that the Book of Genesis (read literally) is a truer account of human origins than one or more competing religious or nonreligious accounts. Thus, the nonestablishment norm *does* forbid government to base political choices on religious arguments in this sense: Government may not base any action—therefore, it may not base any choice, including one about the morality of human conduct—on the view that a religious belief is closer to the truth or otherwise better than one or more competing religious or nonreligious beliefs. (Again, a religious argument is an argument that presupposes the truth of a religious belief and includes that belief as one of its essential premises.) The nonestablishment norm forbids government to base political choices on religious arguments; at least as an ideal matter, the nonestablishment norm requires that if government wants to make a political choice, including one about the morality of human conduct, it do so only on the basis of a secular argument: an argument that relies neither on any religious belief nor on the belief that God does not exist.

As the foregoing discussion suggests, and as I explained earlier in this chapter, the nonestablishment norm also forbids government to base political choices on secular arguments of a certain sort, namely, secular arguments to the effect that one or more religious tenets are more authentically American, or more representative of the sentiments of the community, or otherwise better, than one or more competing religious or nonreligious tenets. When I refer, in describing the requirements of the nonestablishment norm, to a "secular" argument or rationale, I do not mean to include arguments of the sort described in the preceding sentence, but only those that do not in any way valorize one or more religious tenets—that do not claim that one or more religious beliefs are better, along one or another dimension of value, than one or more competing religious or nonreligious beliefs. Again, the central point of the free exercise and nonestablishment norms, taken together, is that government may not make judgments about the value or disvalue—the truth value, the moral value, the social value—of religions or of religious practices or tenets (qua religious).

In making a political choice, especially a political choice about the morality of human conduct, legislators and other public officials sometimes rely *both* on a religious argument *and* on an independent secular argument: a secular argument that, if accepted, supports the choice without help from a religious argument. It is noteworthy, in that regard, that many of those who contend that abortion is immoral, like many of those who contend that homosexual sexual conduct is immoral, come armed with an independent secular argument as well as a religious argument; indeed, some of them come armed only with a secular moral argument. If government based a political choice about the morality of human conduct at least partly on a plausible secular argument that supports the choice, it would be extremely difficult for a court to discern whether government based the choice solely on the secular argument or, instead, only partly on the secular argument and partly on the religious argument. That government would have made the choice even in the absence of the religious argument, solely on the basis of the secular argument, is some evidence that the choice was based solely on the secular argument. Such evidence is not conclusive, however; that one would have made a choice in the absence of reason X does not mean that one did not base the choice on X; it does not even mean that, in making the choice, one did not rely solely on X. Moreover, counterfactual inquiry by a court into whether government "would have made" a political choice about the morality of human conduct in the absence of a religious argument on which some officials relied is so speculative as to be unusually vulnerable to distortion by a judge's own sympathies and hostilities.[90] Indeed, an individual legislator or other public official, inquiring in good faith, might not be able to decide with confidence whether she herself would have made a political choice about the morality of human conduct in the absence of a religious argument on which she relied (or whether she would make it now).

As an ideal matter, the nonestablishment norm is probably best understood, as I have suggested, to forbid government to make any political choice, even one about the morality of human conduct, on the basis of a religious argument. But, given the difficulty emphasized in the preceding paragraph, we should probably conclude that *as a practical matter*, the nonestablishment norm requires only that government not make political choices of the kind in question here—political choices about the morality of human conduct—unless a plausible secular rationale supports the choice without help from a parallel religious argument. (Kathleen Sullivan has written that "the negative bar against establishment of religion implies the affirmative 'establishment' of a civil order for the resolution of public moral disputes. . . . [P]ublic moral disputes may be resolved only on grounds articulable in secular terms."[91]) Under that

approach—which, concededly, involves an "underenforcement" of the full ideal of nonestablishment[92]—a court need not pretend that it can discern what it probably could rarely discern, namely, whether government based such a political choice solely on a secular moral argument or only partly on such an argument and partly on a religious moral argument. (How many legislators would have to base their votes at least partly on a religious moral argument before a court should conclude that "government" had based the political choice on such an argument? How could a court determine whether or not an individual legislator had based her vote on a religious moral argument?) Moreover, if it became known that political choices about the morality of human conduct would be struck down as unconstitutional if not based solely on a secular argument, public officials could, and many doubtless would, take steps to construct a legislative history that would make it even harder for a court to conclude that such a political choice was not based solely on a secular moral argument. The inevitability of such a strategy reinforces the conclusion that as a practical matter, the nonestablishment norm should be understood to require only that government not make a political choice about the morality of human conduct in the absence of a plausible secular rationale.[93] (A qualification is necessary here. I explain, in chapter 3, why a religious argument in support of the claim that each and every human being is sacred presents a special case: Even if we assume that no secular argument supports the claim that every human being is sacred, government may, under the nonestablishment norm, rely on a religious argument in support of the claim.)

Douglas Laycock has dissented from the nonestablishment position I am defending here.[94] According to Laycock, although government may not require anyone to engage in an act of religious worship, government may make a coercive political choice about the morality of human conduct even if the only rationale that supports the choice is religious—that is, even if no plausible secular argument supports the choice.[95] In Laycock's view, government would not violate the nonestablishment norm by outlawing homosexual sexual conduct, for example, or by denying legal recognition to homosexual marriage, even if *ex hypothesi* the only rationale that could support such government action were religious. The absence of a plausible secular rationale is not irrelevant, in Laycock's view, because it might tend to show that government is in fact compelling persons to engage in acts of religious worship. What finally matters for Laycock, however, what is finally determinative for him, is not the presence or absence of a plausible secular rationale but only whether government is compelling anyone to engage in an act of religious worship.

Laycock's position is deeply problematic. True, government may not compel anyone to engage in an act of religious worship. But why may it not do so? Government may not compel anyone to engage in an act of religious worship *because the nonestablishment norm forbids government to impose one or another religion—including one or another understanding of God or of God's will—on anyone.* (Indeed, as I have emphasized in this chapter, government may not make judgments about the value—the truth value, the moral value, the social value—of religions or of religious practices or tenets qua religious.) However, if government (a state legislature, say) makes a coercive political choice about the morality of human conduct, a choice requiring or forbidding persons to do something, and if the only reason or reasons that can support the political choice are religious—if no plausible secular rationale supports the choice—then government has undeniably imposed religion on those persons whom the choice coerces.[96] This is so whether or not the political choice compels persons to engage in what is conventionally understood as an act of religious worship.

Moreover, it is to exalt form over substance to say that under the nonestablishment norm government may not compel anyone to engage in an act of religious worship but that government may make a coercive political choice about the morality of human conduct even if the only reason or reasons that can support the political choice are religious. If I am a gay man or a lesbian, forbidding me to fulfill my sexuality because it is believed that the will of God, as revealed in Leviticus, forbids me to do so is a much more profound and disabling imposition on me—a *religious* imposition—than requiring me, for example, to listen to my public school teacher begin class with a recitation of the Lord's Prayer. (Requiring me to listen to my teacher begin class with a recitation of the Lord's Prayer is a way of compelling me to attend a worship service.)

For reasons having to do both with the central prohibition and meaning of the nonestablishment norm and with the importance of not exalting form over substance, I think Laycock's ground for rejecting my position is quite weak. The nonestablishment rule that Laycock emphasizes, against government compelling anyone to engage in an act of religious worship, is only an instance—albeit, a very important instance—of a more general rule against government imposing one or another religion on anyone. For government to make a coercive political choice about the morality of human conduct that can be supported only by a religious reason or reasons is for government to impose religion.

Admittedly, that under the nonestablishment norm government may not make a political choice about the morality of human conduct unless a plausible secular rationale supports the choice has less practical significance than one might think, because there will be plausible secular rationales for most such political choices that government might want to make. (In adjudicating the constitutionality, under the nonestablishment norm, of a political choice about the morality of human conduct, the proper issue for a court is not whether a secular rationale is, in the court's own view, persuasive. After all, the judiciary does not have the principal policymaking authority or responsibility. The proper issue for a court is only whether a secular rationale is plausible—that is, whether a legislator or other public official could reasonably find the rationale persuasive.[97]) However, that a political choice about the morality of human conduct does not violate the nonestablishment norm does not mean that the choice does not violate some other constitutional requirement. In my view, a state's denial of legal recognition to homosexual marriage probably violates the antidiscrimination part of the Fourteenth Amendment.[98]

I said that under the nonestablishment norm, government may not make a political choice about the morality of human conduct unless a plausible secular rationale supports the choice. But what about an individual legislator or other public official: What should she do? Should she vote to support a political choice about the morality of human conduct if she is agnostic about whether, or even skeptical that, a plausible secular rationale supports the choice, leaving it up to others, and ultimately to the courts, to decide if such a rationale exists? Fidelity to the spirit of the nonestablishment norm seems to me to require more of her: She should vote to support a political choice about the morality of human conduct only if, in her view, a persuasive secular rationale exists. (I am not suggesting that such a constitutional duty could be, or even should be, judicially enforced. How could a court determine whether or not an individual legislator really believes that there is a persuasive secular rationale? The duty would have to be self-enforced.[99]) That she cannot reach a judgment about the soundness of the relevant secular argument or arguments on her own is not disabling, because she can seek the help of those whose judgment she respects and trusts. In chapter 3, I explain why as a matter of political morality, too, and not just as a matter of constitutionality, an individual legislator should vote to support a political choice about the morality of human conduct only if, in her view, a persuasive secular rationale exists.

Constitutional legality does not entail moral propriety; that an act would not violate any constitutional norm does not entail that the act would be, all things considered, morally appropriate. Similarly, constitutional illegality does not entail moral impropriety; that an act would violate a constitutional norm does not entail that the act would be, apart from its unconstitutionality, morally inappropriate.[100] Indeed, if we conclude that an act that would violate a constitutional norm would not be, apart from its unconstitutionality, morally inappropriate—and especially if we conclude that the act would be morally appropriate—we can proceed to inquire whether the constitutional law of the United States shouldn't be revised by the Supreme Court, or even amended pursuant to Article V of the Constitution, to permit the act.[101] Beyond the constitutional inquiry, therefore, lies the moral inquiry.

I have explained that citizens and even legislators and other public officials are constitutionally free to present religious arguments, including religious arguments about the morality of human conduct, in public political debate. The question remains, however, whether, all things considered, it isn't morally inappropriate for citizens and especially legislators and other public officials to present such arguments in public political debate. I turn to that question in chapter 2. According to the construal of the nonestablishment norm I have defended in this chapter, government may rely on a religious argument in making a political choice about the morality of human conduct only if a plausible secular rationale supports the choice. The question remains, however, whether, all things considered, it isn't morally inappropriate for legislators and other public officials, and for citizens voting in a referendum or an initiative election, to rely on a religious argument in making a political choice about the morality of human conduct even if a plausible secular rationale—or even, in their view, a persuasive secular rationale—supports the choice. (I have suggested that those with the principal policymaking authority and responsibility—in particular, legislators—should ask themselves whether they find a secular rationale persuasive.[102]) From the other side, the question remains whether, *apart from the nonestablishment norm*, it isn't morally permissible for legislators and others to rely on a religious argument in making a political choice about the morality of human conduct even if, in their view, no persuasive or even plausible secular rationale supports the choice. I turn to those questions in chapter 3.

Appendix: Declaration on the Elimination of all Forms of Intolerance and of Discrimination Based on Religion or Belief

[In the international law of human rights, the most complete statement of the human right to freedom of religion is the Declaration on the Elimination of All Forms of Intolerance and of Discrimination Based on Religion or Belief, which was adopted without vote by the General Assembly of the United Nations on Nov. 25, 1981 (Resolution 36/55). The Declaration is an elaboration of the position on religious freedom embodied in many other human rights documents, including the Universal Declaration of Human Rights (Articles 18 and 29), the International Covenant on Civil and Political Rights (Article 18), and the European Convention on Human Rights (Article 9).[103] It is instructive to consult the Declaration, therefore—as I have done in this chapter—when evaluating the extent of the freedom of religion protected by the constitutional law of the United States.]

Declaration on the Elimination of All Forms of Intolerance and of Discrimination Based on Religion or Belief, 1981

This Declaration was adopted without vote by the General Assembly of the United Nations on 25 November 1981 (Resolution 36/55). For the background see *Yearbook of the United Nations,* 1981, pp. 879-83. The Declaration was prepared by the Human Rights Commission of the Economic and Social Council. For an earlier proposal of a Draft Convention on the subject see the second edition of this work.

TEXT

The General Assembly

Considering that one of the basic principles of the Charter of the United Nations is that of the dignity and equality inherent in all human beings, and that all Member States have pledged themselves to take joint and separate action in co-operation with the Organization to promote and encourage universal respect for and observance of human rights and fundamental freedoms for all, without distinction as to race, sex, language or religion,

Considering that the Universal Declaration of Human Rights and the International Covenants on Human Rights proclaim the principles of nondiscrimination and equality before the law and the right to freedom of thought, conscience, religion and belief,

Considering that the disregard and infringement of human rights and fundamental freedoms, in particular of the right to freedom of thought, conscience, religion or whatever belief, have brought, directly or indirectly, wars and great suffering to mankind, especially where they serve as a means of foreign interference in the internal affairs of other States and amount to kindling hatred between peoples and nations,

Considering that religion or belief, for anyone who professes either, is one of the fundamental elements in his conception of life and that freedom of religion or belief should be fully respected and guaranteed,

Considering that it is essential to promote understanding, tolerance and respect in matters relating to freedom of religion and belief and to ensure that the use of religion or belief for ends inconsistent with the Charter of the United Nations, other relevant instruments of the United Nations and the purposes and principles of the present Declaration is inadmissible,

Convinced that freedom of religion and belief should also contribute to the attainment of the goals of world peace, social justice and friendship among peoples and to the elimination of ideologies or practices of colonialism and racial discrimination,

Noting with satisfaction the adoption of several, and the coming into force of some, conventions, under the aegis of the United Nations and of the specialized agencies, for the elimination of various forms of discrimination,

Concerned by manifestations of intolerance and by the existence of discrimination in matters of religion or belief still in evidence in some areas of the world,

Resolved to adopt all necessary measures for the speedy elimination of such intolerance in all its forms and manifestations and to prevent and combat discrimination on the ground of religion or belief,

Proclaims this Declaration on the Elimination of All Forms of Intolerance and of Discrimination Based on Religion or Belief:

Article 1

1. Everyone shall have the right to freedom of thought, conscience and religion. This right shall include freedom to have a religion or whatever belief of his choice, and freedom, either individually or in community with others and in public or private, to manifest his religion or belief in worship, observance, practice and teaching.

2. No one shall be subject to coercion which would impair his freedom to have a religion or belief of his choice.

3. Freedom to manifest one's religion or beliefs may be subject only to such limitations as are prescribed by law and are necessary to protect public safety, order, health or morals or the fundamental rights and freedoms of others.

Article 2

1. No one shall be subject to discrimination by any State, institution, group of persons, or person on grounds of religion or other beliefs.

2. For the purposes of the present Declaration, the expression "intolerance and discrimination based on religion or belief" means any distinction, exclusion, restriction or preference based on religion or belief and having as its purpose or as its effect nullification or impairment of the recognition, enjoyment or exercise of human rights and fundamental freedoms on an equal basis.

Article 3

Discrimination between human beings on grounds of religion or belief constitutes an affront to human dignity and a disavowal of the principles of the Charter of the United Nations, and shall be condemned as a violation of the human rights and fundamental freedoms proclaimed in the

Universal Declaration of Human Rights and enunciated in detail in the International Covenants on Human Rights, and as an obstacle to friendly and peaceful relations between nations.

Article 4

1. All States shall take effective measures to prevent and eliminate discrimination on the grounds of religion or belief in the recognition, exercise and enjoyment of human rights and fundamental freedoms in all fields of civil, economic, political, social and cultural life.

2. All States shall make all efforts to enact or rescind legislation where necessary to prohibit any such discrimination, and to take all appropriate measures to combat intolerance on the grounds of religion or other beliefs in this matter.

Article 5

1. The parents or, as the case may be, the legal guardians of the child have the right to organize the life within the family in accordance with their religion or belief and bearing in mind the moral education in which they believe the child should be brought up.

2. Every child shall enjoy the right to have access to education in the matter of religion or belief in accordance with the wishes of his parents or, as the case may be, legal guardians, and shall not be compelled to receive teaching on religion or belief against the wishes of his parents or legal guardians, the best interests of the child being the guiding principle.

3. The child shall be protected from any form of discrimination on the ground of religion or belief. He shall be brought up in a spirit of understanding, tolerance, friendship among peoples, peace and universal brotherhood, respect for freedom of religion or belief of others, and in full consciousness that his energy and talents should be devoted to the service of his fellow men.

4. In the case of a child who is not under the care either of his parents or of legal guardians, due account shall be taken of their expressed wishes or of any other proof of their wishes in the matter of religion or belief, the best interests of the child being the guiding principle.

5. Practices of a religion or beliefs in which a child is brought up must not be injurious to his physical or mental health or to his full development, taking into account Article I, paragraph 3, of the present Declaration.

Article 6

In accordance with Article I of the present Declaration, and subject to the provisions of Article 1, paragraph 3, the right to freedom of thought, conscience, religion or belief shall include, *inter alia,* the following freedoms:

(*a*) To worship or assemble in connection with a religion or belief, and to establish and maintain places for these purposes;
(*b*) To establish and maintain appropriate charitable or humanitarian institutions;
(*c*) To make, acquire and use to an adequate extent the necessary articles and materials related to the rites or customs of a religion or belief;
(*d*) To write, issue and disseminate relevant publications in these areas;
(*e*) To teach a religion or belief in places suitable for these purposes;
(*f*) To solicit and receive voluntary financial and other contributions from individuals and institutions;
(*g*) To train, appoint, elect or designate by succession appropriate leaders called for by the requirements and standards of any religion or belief;
(*h*) To observe days of rest and to celebrate holidays and ceremonies in accordance with the precepts of one's religion or belief;
(*i*) To establish and maintain communications with individuals and communities in matters of religion and belief at the national and international levels.

Article 7
The rights and freedoms set forth in the present Declaration shall be accorded in national legislation in such a manner that everyone shall be able to avail himself of such rights and freedoms in practice.

Article 8
Nothing in the present Declaration shall be construed as restricting or derogating from any right defined in the Universal Declaration of Human Rights and the International Covenants on Human Rights.

Chapter Two

RELIGIOUS ARGUMENTS IN PUBLIC POLITICAL DEBATE

The proper role of religious arguments in politics is the subject of intense controversy in the United States. The controversy comprises two inquiries: a debate about the *constitutionally* proper role of such arguments in politics and a related but distinct debate about their *morally* proper role. In the preceding chapter, I addressed the question of the constitutionally proper (permissible) role of religious arguments in American politics; in this and the next chapter, I address the question of the morally proper role of such arguments in politics. Because it focused on the constitutional law of the United States, the preceding chapter was relevant to questions about "religion in politics" as they arise in the United States. This and the next chapter, however, are relevant to such questions as they arise in any democratic political community that, like the United States, is religiously pluralistic.

By "morally" proper role, I mean simply the role that, taking into account every relevant consideration (other than constitutionality), we should deem it permissible or proper for religious arguments to play in politics. In this chapter, the "we" refers principally (though not exclusively) to those religious nonbelievers who are skeptical about welcoming religiously-based moral discourse into the political discourse of a liberal democracy like the United States; the argument of this chapter is addressed principally to such nonbelievers. By contrast, the principal argument of the next chapter is addressed principally to religious believers; in the next chapter, the "we" refers principally to those religious believers—in particular, to those Christians—who are skeptical about accepting any limits on religiously-based moral arguments as a basis of political choice. (Christians remain the largest religious group in the United States.)

I concluded, in chapter 1, that citizens and even legislators and other public officials are constitutionally free to present religious arguments, including religious arguments about the morality of human conduct, in public debate about what political choices to make. (To repeat, the political choices with which I am principally concerned in this book are political choices about the morality of human conduct: choices to ban or otherwise disfavor one or another sort of human conduct on the basis of the view that it is immoral for human beings, whether all or some, to engage in the conduct.) I conclude, in this chapter, that as a matter not only of constitutionality but also of political morality, citizens and even legislators and other public officials may present, in public political debate, religious arguments about the morality of human conduct. Indeed, I conclude that it is important that such religious arguments, no less than secular arguments about the morality of human conduct, be presented in public political debate. It bears emphasis that the inquiry I pursue in this chapter, about the role of religious arguments in public political debate, is about political morality, not political strategy. "[T]he distinction between principle and prudence should be emphasized. The fundamental question is not whether, as a matter of prudent judgment in a religiously pluralist society, those who hold particular religious views ought to cast their arguments in secular terms. Even an outsider can say that the answer to that question is clearly, 'Yes, most of the time,' for only such a course is likely to be successful overall."[1]

I. Religious Arguments in Public Political Debate—and in Public Culture Generally

I explained, in chapter 1, that citizens and even legislators (and other public officials) are constitutionally free to present religious arguments about the morality of human conduct in public political debate. Even so, should such arguments be presented in public political debate? Again, that one is constitutionally free to do something does not mean that as a matter of morality one should do it; it does not mean that it is morally appropriate for one to do it. Constitutional legality no more entails moral propriety than constitutional illegality entails moral impropriety.

It is inevitable that some legislators, and some citizens participating in a referendum or an initiative election, will rely on—will put at least some weight on—religious arguments in voting for political choices about the morality of human conduct. Moreover, a religious argument can be quite influential in moving a critical mass of legislators or citizens to want to make a particular political choice and in inclining them to accept, as a rationale for the choice, a secular argument that supports the choice. For example, a biblically based argument that homosexual sexual conduct is immoral[2] has moved some citizens and legislators to want to deny legal recognition to homosexual marriage and has inclined some of them to accept, as a secular rationale for their position, the argument that homosexuality, like alcoholism, is pathological and ought not to be indulged, or the argument that recognizing homosexual marriage would threaten the institution of heterosexual marriage and other "traditional family values". Because of the role that religiously based moral arguments inevitably play in the political process, then, it is important that such arguments, no less than secular moral arguments, be presented in, so that they can be tested in, public political debate. Ideally, such arguments will sometimes be tested, in the to and fro of public political debate, by competing scripture- or tradition-based religious arguments. Luke Timothy Johnson's warning is relevant here:

> If liberal Christians committed to sexual equality and religious tolerance abandon these texts as useless, they also abandon the field of Christian hermeneutics to those whose fearful and—it must be said—sometimes hate-filled apprehension of Christianity will lead them to exploit and emphasize just those elements of the tradition that have proved harmful to humans. If what Phyllis Trible has perceptively termed "texts of terror" within the Bible are not encountered publicly and engaged intellectually by a hermeneutics that is at once faithful and critical, then they will continue to exercise their potential for harm among those who, without challenge, can claim scriptural authority for their own dark impulses.[3]

Nonetheless, some persons want to keep religiously based moral arguments out of public political debate as much as possible. For example, American philosopher Richard Rorty has written approvingly of "privatizing religion—keeping it out of . . . 'the public square,' making it seem bad taste to bring religion into discussions of public policy."[4] One reason for wanting to "privatize" religion is that religious debates about controversial political issues can be quite divisive. But American history does not suggest that religious debates about controversial issues—racial discrimination, for example, or war—are invariably more divisive than secular debates about those or other issues.[5] Some issues are so controversial that debate about them is inevitably divisive without regard to whether the debate is partly religious or, instead, only secular.[6]

Another reason for wanting to keep religiously based moral arguments out of public political debate focuses on the inability of some persons to gain a critical distance on their religious beliefs—the kind of critical distance essential to truly deliberative debate. But in the United States and in other liberal democracies, many persons *are* able to gain a critical distance on their religious beliefs;[7] they are certainly as able to do so as they and others are able to gain a critical distance on other fundamental beliefs.[8] David Tracy speaks for many of us religious believers when he writes:

> For believers to be unable to learn from secular feminists on the patriarchal nature of most religions or to be unwilling to be challenged by Feuerbach, Darwin, Marx, Freud, or Nietzsche is to refuse to take seriously the religion's own suspicions on the existence of those fundamental distortions named sin, ignorance, or illusion. The interpretations of believers will, of course, be grounded in some fundamental trust in, and loyalty to, the Ultimate Reality both disclosed and concealed in one's own religious tradition. But fundamental trust, as any experience of friendship can teach, is not immune to either criticism or suspicion. A religious person will ordinarily fashion some hermeneutics of trust, even one of friendship and love, for the religious classics of her or his tradition. But, as any genuine understanding of friendship shows, friendship often demands both critique and suspicion. A belief in a pure and innocent love is one of the less happy inventions of the romantics. A friendship that never includes critique and even, when appropriate, suspicion is a friendship barely removed from the polite and wary communication of strangers. As Buber showed, in every I-Thou encounter, however transient, we encounter some new dimension of reality. But if that encounter is to prove more than transitory, the difficult ways of friendship need a trust powerful enough to risk itself in critique and suspicion. To claim that this may be true of all our other loves but not true of our love for, and trust in, our religious tradition makes very little sense either hermeneutically or religiously.[9]

Of course, it is undeniable that some religious believers are unable to gain much if any critical distance on their fundamental religious beliefs. As so much in the twentieth century attests, however, one need not be a religious believer to adhere to one's fundamental beliefs with closeminded or even fanatical tenacity.

Although no one who has lived through recent American history can believe that religious contributions to the public discussion of difficult moral issues are invariably deliberative rather than dogmatic, there is no reason to believe that religious contributions are never deliberative. Religious discourse about the difficult moral issues that engage and divide us citizens of liberal democratic societies is not necessarily more problematic—more monologic, say—than resolutely secular discourse about those issues. Because of the religious illiteracy—and, alas, even prejudice—rampant among many nonreligious intellectuals,[10] we probably need reminding that, at its best, religious discourse in public culture is not less dialogic—not less openminded, not less deliberative—than is, at its best, secular discourse in public culture. (Nor, at its worst, is religious discourse more monologic—more closeminded and dogmatic—than is, at its worst, secular discourse.)[11] David Hollenbach's work has developed this important point: "Much discussion of the public role of religion in recent political thought presupposes that religion is more likely to fan the flames of discord than contribute to social concord. This is certainly true of some forms of religious belief, but hardly of all. Many religious communities recognize that their traditions are dynamic and that their understandings of God are not identical with the reality of God. Such communities have in the past and can in the future engage in the religious equivalent of intellectual solidarity, often called ecumenical or interreligious dialogue."[12]

A central feature of Hollenbach's work is his argument, which I accept, that the proper role of "public" religious discourse in a society as religiously pluralistic as the United States is a role to be played, in the main, much more in public culture—in particular, "in those components of civil society that are the primary bearers of cultural meaning and value—universities, religious communities, the world of the arts, and serious journalism"—than in public debate specifically about political issues.[13] He writes: "[T]he domains of government and policy-formation are not generally the appropriate ones in which to argue controverted theological and philosophical issues. . . ."[14] But, as Hollenbach goes on to acknowledge, "it is nevertheless neither possible nor desirable to construct an airtight barrier between politics and culture."[15]

There is, then, in addition to the reasons I have already given, this important reason for not opposing the presentation of religiously based moral arguments in public political debate: In a society as overwhelmingly religious as the United States, we do present and discuss—and we should present and discuss—religiously based moral arguments in our public culture.[16] Rather than try to do the impossible—maintain a wall of separation ("an airtight barrier") between the religiously based moral discourse that inevitably and properly takes place in public culture ("universities, religious communities, the world of the arts, and serious journalism") on the one side and the discourse that takes place in public political debate ("the domains of government and policy-formation") on the other side—we should simply welcome the presentation of religiously based moral arguments in *all* areas of our public culture, *including* public debate specifically about contested political choices.[17] Indeed, for the reasons I have given, we should not merely welcome but *encourage* the presentation of such arguments in public politicial debate—so that we can test them there.

To be sure, religious discourse in public—whether in public political debate or in other parts of our public culture—is sometimes quite sectarian and therefore divisive. But religiously based moral discourse is not necessarily more sectarian than secular moral discourse. It can be much less sectarian. After all, certain basic moral premises common to the Jewish and Christian traditions, in conjunction with the supporting religious premises, still constitute the fundamental moral horizon of most Americans—much more so than do Kantian (or neo-Kantian) premises, or Millian premises, or Nietzschean premises, and so forth.[18] According to John Coleman, "the tradition of biblical religion is arguably the most powerful and pervasive symbolic resource" for public ethics in the United States today. "[O]ur tradition of religious ethics seems . . . to enjoy a more obvious public vigor and availability as a resource for renewal in American culture than either the tradition of classic republican theory or the American tradition of public philosophy." Coleman reminds us that "the strongest American voices for a compassionate just community always appealed in public to religious imagery and sentiments, from Winthrop and Sam Adams, Melville and the Lincoln of the second inaugural address, to Walter Rauschenbusch and Reinhold Niebuhr and Frederick Douglass and Martin Luther King." As Coleman explains, "The American religious ethic and rhetoric contain rich, polyvalent symbolic power to command sentiments of emotional depth, when compared to 'secular' language, . . . [which] remains exceedingly 'thin' as a symbol system." Coleman emphasizes that "when used as a public discourse, the language of biblical religion is beyond the control of any particular, denominational theology. It represents a common American cultural patrimony. . . . American public theology or religious ethics . . . cannot be purely sectarian. The biblical language belongs to no one church, denomination, or sect." In Coleman's view, "The genius of public American theology . . . is that it has transcended denominations, been espoused by people as diverse as Abraham Lincoln and Robert Bellah who neither were professional theologians nor belonged to any specific church and, even in the work of specifically trained professional theologians, such as Reinhold Neibuhr, has appealed less to revelational warrant for its authority within public policy discussions than to the ability of biblical insights and symbols to convey a deeper human wisdom. . . . Biblical imagery . . . lies at the heart of the American self-understanding. It is neither parochial nor extrinsic."[19]

So, religiously based moral discourse is not always more sectarian than secular moral discourse; it can be less sectarian. But even if religiously based moral discourse were invariably more sectarian than secular moral discourse, this important point would remain: Sectarian discourse, including sectarian religious discourse, can make a worthwhile contribution to public deliberation about difficult moral issues. As Jeremy Waldron has explained:

> Even if people are exposed in argument to ideas over which they are bound to disagree—and how could *any* doctrine of public deliberation preclude *that*?—it does not follow that such exposure is pointless or oppressive. For one thing, it is important for people to be acquainted with the views that others hold. Even more important, however, is the possibility that my own view may be improved, in its subtlety and depth, by exposure to a religion or a metaphysics that I am initially inclined to reject. . . . I mean . . . to draw attention to an experience we all have at one time or another, of having argued with someone whose world view was quite at odds with our own, and of having come away thinking, "I'm sure he's wrong, and I can't follow much of it, but, still, it makes you think . . .". The prospect of losing that sort of effect in public discourse is, frankly, frightening—terrifying, even, if we are to imagine it being replaced by a kind of "deliberation" that, in the name of "fairness" or "reasonableness" (or worse still, "balance") consists of bland appeals to harmless nostrums that are accepted without question on all sides. This is to imagine open-ended public debate reduced to the formal trivia of American televisions networks. . . . [This] might apply to *any* religious or other philosophically contentious intervention. We do not have (and we should not have) so secure a notion of public consensus, or such stringent requirements of fairness in debate, as to exclude any view from having its effect in the marketplace of ideas.[20]

Again, Richard Rorty thinks that it makes sense to "privatiz[e] religion—[to] keep[ ] it out of . . . 'the public square,' making it seem bad taste to bring religion into discussions of public policy."[21] Rorty should think again. Not only are the reasons for wanting to privatize religion weak, there are strong countervailing reasons, which I have given in this section, for wanting to "public-ize" religion, not privatize it. We should welcome religiously based moral arguments into the public square (where we can then test them), not try to keep them out. We should make it seem bad taste to sneer when people bring their religious convictions to bear in public discussions of controversial political issues, like homosexuality and abortion. It is not *that* religious convictions are brought to bear in public political debate that should worry us, but *how* they are sometimes brought to bear (e.g., dogmatically). But we should be no less worried about how fundamental secular convictions are sometimes brought to bear in public political debate.[22]

II. Greenawalt on Religious Arguments in Public Political Debate

Kent Greenawalt is one of the most thoughtful contributors to the debate about the morally proper role of religion in politics. By way of defending the position I have been presenting in this chapter, I want to explain why I disagree with Greenawalt's position that legislators should not present religious arguments in public political debate.[23] (Greenawalt writes: "I concentrate on legislators, assuming that chief executives [the President of the United States and state governors] should be governed by similar standards."[24]) In the preceding section, I explained why we should encourage the airing of religious arguments, even by legislators, in public political debate. Now I want to explain why Greenawalt's reasons for asking legislators to avoid presenting such arguments in public political debate do not bear the weight he puts on them.

Greenawalt's rationale is twofold. His first and principal reason is that if a legislator presents a religious argument in public political debate, some of those the legislator represents "are likely to feel imposed upon in the sense of being excluded . . . ."[25] Greenawalt is concerned about "the inequality and disrespect that members of minorities may feel . . . , the sense they may have that they are being imposed upon as second class citizens."[26] In my view, this reason will not bear the weight Greenawalt puts on it. Unless my representative and I are clones, there will almost certainly be some occasions, perhaps many, on which she and I are in fundamental disagreement. Why should I feel significantly more imposed upon, if imposed upon at all, if our disagreement is rooted in religious differences than if it is rooted in secular differences? In a different but related context, Steven Smith has written something that is relevant here—and I concur in it:

> [T]he very concept of "alienation," or symbolic exclusion, is difficult to grasp. How, if at all, does "alienation" differ from "anger," "annoyance," "frustration," or "disappointment" that every person who finds himself in a political minority is likely to feel? "Alienation" might refer to nothing more than an awareness by an individual that she belongs to a religious minority, accompanied by a realization that at least on some issues she is unlikely to be able to prevail in the political process. . . . That awareness may be discomforting. But is it the sort of phenomenon for which constitutional law can provide an efficacious remedy? Constitutional doctrine that stifles the message will not likely alter the reality—or a minority's awareness of that reality.[27]

Continuing to develop his first reason, Greenawalt writes: "[A]t least for many religious arguments, the speaker seems to put himself or herself in a kind of privileged position, as the holder of a *basic* truth that many others lack. This assertion of privileged knowledge may appear to imply inequality of status that is in serious tension with the fundamental idea of equality of citizens within liberal democracies."[28] It is true for many secular arguments, too, however, that the speaker seems to portray herself as the holder of basic truths or insights that many others lack—for example, basic truths about human nature, or about the workings of society. In any event, I fail to see how such an "assertion of privileged knowledge" is in any way inconsistent with the fundamental idea of the equality of all citizens. This idea of equality is really two, related ideas:

- The idea of the *moral* equality of all persons: Every person, without regard to "race, colour, sex, language, religion, political or other opinion, national or social origin, birth or other status,"[29] is sacred.
- The idea of the *political* equality of all citizens: Because every person is sacred, every citizen is sacred, and therefore every citizen, without regard to race, sex, religion, etc., is entitled to participate in the politics and government of his or her society on an equal basis with every other citizen; moreover, no citizen may be treated with less respect or concern than any other citizen.

This twofold idea of the fundamental equality of all citizens entails neither that all citizens are in the grip of the same convictions, religious or not, nor that all their different convictions are (or that they should act as if they are) equally correct. For a legislator to present a religious argument in public political debate is not necessarily for her to assert, imply, or presuppose a denial of the fundamental equality of all citizens. (This is not to deny that one of her convictions, religious or not—for example, the conviction that only white persons are entitled to vote—may itself assert, imply, or presuppose a denial of the fundamental equality of all citizens.)

According to Greenawalt, "[w]hen legislators speak on political issues, they represent all their constituents. Their explicit reliance on any controversial religious or other comprehensive view would be inappropriate."[30] Yes, legislators should represent *all* their constituents. But the sense in which they should do so is that in making political choices, legislators should be concerned with the good or well-being of all their constituents. Indeed, ideally a legislator should aim, to the extent possible, at the good of every member of the political community, rather than at the good merely of some members—or, worse, at the good merely of the legislator herself.[31] *That* is the real sense in which legislators should represent all their constituents. It is not necessarily inconsistent with the duty of legislators to represent all their constituents for a legislator, in speaking on political issues, to invoke a "controversial religious or other comprehensive view"—for example, a religious belief about what is truly good for every member of the community.[32]

Moreover, it is virtually axiomatic that in a liberal democratic society the truthful disclosure of all the reasons why one's representative is inclined to stand where she does is an overriding, if infrequently honored, value. I suspect that most of us citizens of a liberal democracy would be more than willing to endure some feeling of being "imposed upon" if that were the price for knowing all the reasons why our representatives stand where they do. Among other things, if we know all the reasons, we can respond more effectively—especially when our representatives are up for reelection—than if we know only some of them.

Especially because the truthful disclosure of all the reasons why an elected official stands where she does is an overriding value, a much more sensible way to minimize the extent to which *some* citizens might "feel imposed upon in the sense of being excluded"—a much more proportionate way to minimize "the inequality and disrespect that members of minorities may feel . . . , the sense that they are being imposed upon as second class citizens"—is for a legislator to feature, in public political debate, not only all the relevant arguments that she takes seriously, including religious arguments, but *all* the credible and not otherwise inappropriate arguments that might incline a citizen to support the political choice at issue. In that way, a legislator does not conceal the real bases of her support, but neither does she gratuitously marginalize or exclude reasons that might appeal to some of her constituents or to some citizens generally; instead, she "re-presents" all the relevant reasons, both those that are most important to her and those that might be most important to someone else. She thereby cultivates the bonds of political community even as she forthrightly indicates why she stands where she does. Obviously, strategic considerations should give any elected official ample incentive to do just what I recommend here.

Greenawalt's second basic reason for concluding that legislators should not present religious arguments in public political debate presupposes his first: Because a legislator's presentation of a religious argument will make citizens who reject the belief "feel imposed upon in the sense of being excluded", the position that legislators may present such arguments in public political debate "underestimates the harm of a religious politics in the present United States."[33] Greenawalt allows that "[i]ntense religious politics in the United States probably would not produce extensive outright violence, but we are still far from harmonious mutual respect and tolerance. Religious divisions are still very significant in many regions, and people are acutely conscious of whether they are in a majority or minority."[34] Greenawalt's estimate of the likely "harm of a religious politics in the United States" is exaggerated. However one evaluates the phenomenon, religion has been, for the most part, domesticated in western liberal democratic societies.[35] As Greenawalt himself emphasizes, "[t]here is a long history of religious involvement in politics in the United States, and most crucial facts are beyond dispute."[36] One crucial fact, beyond dispute, is that notwithstanding that long history of religious involvement in politics in the United States, the sky has not fallen. Indeed, "the risk of major instability generated by religious conflict is minimal. Conditions in modern democracies may be so far from the conditions that gave rise to the religious wars of the sixteenth century that we no longer need worry about religious divisiveness as a source of substantial social conflict."[37] Moreover, even if legislators may present religious arguments in public political debate, the fact remains that in a religiously pluralistic society like the United States, strategic considerations give politicians a powerful incentive to feature—to give pride of place to—secular arguments, if not indeed to present only secular arguments. All the more reason, then, why Greenawalt's fears about what will ensue if we countenance legislators presenting religious arguments in public political debate seem exaggerated: That they *may* feature such arguments does not mean that they *will* do so very often.

Note, too, that Greenawalt accepts David Hollenbach's argument that, quite apart from public argument specifically about political issues, there is an important place for religious discourse in the public culture of the United States.[38] Indeed, Greenawalt recommends that "elected officials [not] actually conceal the most fundamental ground of their convictions, either when in office or running for office. In this respect, I think Jimmy Carter was an apt example. In contrast to John Kennedy, he did not assert that his religious beliefs were irrelevant to his political functioning, and he made clear his deep Protestant evangelical beliefs."[39] It is difficult to understand, then, why Greenawalt believes that an elected official's presentation of a religious argument in public political debate poses dangers not already posed—why, for example, it makes one who rejects the religious premise (or premises) of the argument feel more "imposed upon in the sense of being excluded" than she already does (if she does)—given that, if Greenawalt's recommendation is being followed, the official's embrace of the religious premise comes as no surprise to anyone.

Neither of Greenawalt's two principal arguments against legislators presenting religious arguments in public political debate is persuasive.Greenawalt also argues that the presentation of religious arguments by legislators will not contribute much if anything to the deliberative quality of public political debate.[40] But, as I explained in the preceding section, secular arguments do not always fare better than religious arguments in that regard. It seems doubtful, in any event, that the airing of religious arguments by legislators is likely to so compromise the existing dialogic quality of public political debate—which, of course, is already often depressingly low—that we should want our representatives to be less than fully forthcoming about why they stand where they do.[41] Because our representatives should be fully forthcoming about why they stand where they do, and for the other reasons presented in this chapter, we should encourage the airing of religious arguments, even by legislators, in public political debate about the morality of human conduct.

III. Rawls's "Ideal of Public Reason"

Because it represents a prominent and influential position different from the one I have defended in this chapter, I now want to examine "the idea of public reason" John Rawls has espoused in his book, *Political Liberalism*,[42] The idea of public reason—or, as Rawls often puts it, "the *ideal* of public reason"[43]—is meant by Rawls to regulate, to govern, certain aspects of the politics of a society, like the United States, committed to political liberalism. In Rawls's view, the ideal of public reason is a constituent of political liberalism, and a commitment to the latter should therefore include a commitment to the former.[44]

The important distinction for Rawls is not between religious beliefs or reasons and secular reasons, but between public reasons and nonpublic reasons:

> [T]here are many nonpublic reasons. . . . Among the nonpublic reasons are those of associations of all kinds: churches and universities, scientific societies and professional groups. . . . [T]o act reasonably and responsibly, corporate bodies, as well as individuals, need a way of reasoning about what is to be done. This way of reasoning is public with respect to their members, but nonpublic with respect to political society and to citizens generally. Nonpublic reasons comprise the many reasons of civil society and belong to what I have called the "background culture," in contrast with the political culture.[45]

Although, then, religious reasons are not, for Rawls, the only nonpublic reasons, they are a paradigmatic example of nonpublic reasons. In limiting the political role of nonpublic reasons, Rawls's ideal of public reason limits the political role of religious reasons (among other nonpublic reasons).

Rawls means the ideal to govern not "all political questions but only . . . those involving what we may call 'constitutional essentials' and questions of basic justice." Rawls gives, as examples of "such fundamental questions", "who has the right to vote, or what religions are to be tolerated, or who is to be assured fair equality of opportunity, or to hold property. These and similar questions are the special subject of public reason."[46] However, Rawls limits the jurisdiction of the ideal of public reason to such questions only provisionally: "[M]y aim is to consider first the strongest case where the political questions concern the most fundamental matters. If we should not honor the limits of public reason here, it would seem we need not honor them anywhere. Should they hold here, we can then proceed to other cases. Still, I grant that it is usually highly desirable to settle political questions by invoking the values of public reason."[47] I shall discuss the ideal of public reason, therefore, as if it applied to political questions beyond just "'constitutional essentials' and questions of basic justice". (Larry Solum, who is sympathetic to Rawls's project, has argued that the ideal of public reason should not be limited to the resolution of what Rawls calls "political questions concern[ing] the most fundamental matters", but should "extend . . . to all coercive uses of state power."[48])

Rawls means the ideal of public reason to govern not only the elected representatives of the people (and judges[49]), but all citizens.[50] According to the ideal, neither citizens nor their representatives should make a political choice unless it can be justified on the basis of public reasons, and the public justification of a political choice should be on the basis of public reasons.[51] My principal concern in this chapter is the role of religious arguments in public political debate. (In the next chapter, I turn to the question of the role of such arguments as basis of political choice.) Nonetheless, because Rawls means the ideal of public reason to govern the making as well as the public justification of political choices, I comment here on the ideal as it applies both to the making of and to the public justification of political choices.

Of course, Rawls means for the ideal of public reason to constrain citizens and their representatives, both in making and in publicly justifying political choices, to rely on public, rather than on nonpublic, reasons. But what, according to Rawls, are public reasons? Although Rawls does not distinguish between normative and nonnormative reasons in developing the ideal of public reason, it is, I think, useful to do so. Nonnormative reasons or premises are claims of fact, claims about the way things are (or were, or will be). Normative premises are claims of value, claims about the way things should be. (Although many religious premises are normative, many others are nonnormative; many religious premises are claims about the ways things are.) With respect to nonnormative premises, public reasons are "the plain truths now widely accepted, or available, to citizens generally."[52] Rawls includes here the "conclusions of science when these are not controversial."[53] With respect to normative premises, public reasons are "the ideals and principles expressed by society's conception of political justice . . ."[54] Finally, public reasons include "guidelines of inquiry: principles or reasoning and rules of evidence in the light of which citizens are to decide whether substantive principles properly apply and to identify laws and policies that best satisfy them."[55] Rawls mentions, in particular, "forms of reasoning found in common sense, and the methods . . . of science when these are not controversial."[56]

Rawls does not suppose—and it is obviously not the case—that a consensus has been achieved in the United States about what the constituents of a conception of political justice are. What does Rawls mean, then, by saying that (on the normative side) public reasons are limited to "the ideals and principles expressed by society's conception of political justice"?

> The point of the ideal of public reason is that citizens are to conduct their fundamental discussions within the framework of what each regards as a political conception of justice based on values that the others can reasonably be expected to endorse and each is, in good faith, prepared to defend that conception so understood. This means that each of us must have, and be ready to explain, a criterion of what principles and guidelines we think other citizens (who are also free and equal) may reasonably be expected to endorse along with us. We must have some test we are ready to state as to when this condition is met. . . .
>
> Of course, we may find that actually others fail to endorse the principles and guidelines our criterion selects. This is to be expected. The idea is that we must have such a criterion and this alone already imposes very considerable discipline on public discussion. Not any value is reasonably said to meet this test, or to be a political value; and not any balance of political values is reasonable. It is inevitable and often desirable that citizens have different views as to the most appropriate political conception [of justice]; for the public political culture is bound to contain different fundamental ideas that can be developed in different ways. An orderly contest between them over time is a reliable way to find which one, if any, is most reasonable.[57]

Rawls's point, then, seems to be this: The ideal of public reason constrains a citizen to rely on, on the nonnormative side, only "the plain truths now widely accepted, or available, to citizens generally" and, on the normative side, only those normative premises (ideals, principles, values) that are part of a conception of political justice that (1) she reasonably believes other (free and equal) citizens reasonably could accept and (2) she is prepared to defend to other citizens as one they reasonably could accept.[58] (The ideal of public reason also constrains a citizen to accept certain "guidelines of inquiry . . . [e.g.,] forms of reasoning found in common sense, and the methods . . . of science when these are not controversial.")

It seems quite uncontroversial that if a citizen can justify a political choice on the basis of premises, normative or nonnormative or both, that she believes other citizens reasonably could accept and that she is prepared to defend to them as premises they could accept, she should do so. Strategically, she is much better off doing so than relying on premises she believes other citizens could not accept. Morally, she is much better off doing so, in this sense: She cultivates rather than subverts the bonds of political community (or, as Rawls prefers, "social unity"[59]) by relying on premises that in her view unite, or could unite, the citizenry rather than on premises that divide them. (I have discussed the nature of political community, understood as a "community of judgment", elsewhere—and I have explained why political community, thus understood, is a good.[60])

But, of course, it is not necessarily the case that a citizen can justify a political choice she wants to make—and perhaps even believes herself morally obligated to make—on the basis of premises that she believes other citizens could reasonably accept (and that she is prepared to defend to them as premises they could accept): The relevant premises (that she believes other citizens could reasonably accept) might be indeterminate—or, more precisely, underdeterminate;[61] they might well be inconclusive with respect to the issue at hand. What is she to do when the relevant premises—and therefore the political conception of justice they constitute—are underdeterminate? If Rawls believes that such underdeterminacy is a marginal reality and therefore a minor problem, as in a recent writing he suggests,[62] he is simply wrong. As American constitutional materials powerfully confirm, the underdeterminacy of some of our fundamental political-moral norms is a central reality.[63] In his presidential address to the American Philosophical Association, Philip Quinn has emphasized, with particular reference to the matter of abortion, the problem of undeterminacy that confronts Rawls's ideal of public reason.[64]

But even if for the sake of argument we put aside the problem of underdeterminacy, which is a substantial problem for Rawls's position, this more fundamental question remains: If the premises that a citizen believes other citizens could reasonably accept (and that she is prepared to defend to them as premises they could accept) do not support a political choice she wants to make, why shouldn't she defend the choice—and make it—on the basis of nonpublic premises that in her view do support the choice? Imagine, for example, that because of one or more of her religious beliefs—which, for Rawls, are nonpublic reasons—a citizen is convinced that a political choice is the right political choice, the correct one, but that she also believes (perhaps mistakenly) that the choice is not one she can justify on the basis of public reasons. (Religious premises might be nonnormative—they might be beliefs about the ways things are—as well as normative.) Why should she acquiesce in Rawls's view that the ideal of public reason trumps what she is convinced to be the right political choice? Or, as Rawls himself has an imaginary interlocutor state the question: "[W]hy should citizens in discussing and voting on the most fundamental political questions honor the limits of public reason? How can it be either reasonable or rational, when basic matters are at stake, for citizens to appeal only to a public conception of justice and not to the whole truth as they see it? Surely, the most fundamental questions should be settled by appealing to the most important truths, yet these may far transcend public reason!"[65]

Rawls responds to this, the most fundamental question we can ask about his ideal of public reason, by invoking what he calls "the liberal principle of legitimacy", according to which not even a majority of citizens may exercise coercive political power over any citizen unless a premise or premises that the citizen, understood as free and equal, could reasonably accept supports the majority's doing so.

> [W]hen may citizens by their vote properly exercise their coercive political power over one another when fundamental questions are at stake? Or in the light of what principles and ideals must we exercise that power if our doing so is to be justifiable to others [understood] as free and equal [citizens]? To this question political liberalism replies: our exercise of political power is proper and hence justifiable only when it is exercised in accordance with a constitution the essentials of which all citizens may reasonably be expected to endorse in the light of principles and ideals acceptable to them as reasonable and rational. This is the liberal principle of legitimacy. And since the exercise of political power itself must be legitimate, the ideal of citizenship imposes a moral, not a legal, duty—the duty of civility—to be able to explain to one another on those fundamental questions how the principles and policies they advocate and vote for can be supported by the political values of public reason.[66]

The problem with this response—a conspicuous problem, in my view—is that it is question-begging: Rawls's answer presupposes the authority of that which is at issue. The question why a majority of citizens (we're talking about a democracy, after all) should abandon the coercive political choice they believe they should otherwise make just because, in their view, no premise that other citizens could (reasonably) accept supports the choice *is* the question why a majority of citizens may exercise coercive political power over a citizen only if a premise or premises that they believe the citizen (understood as free and equal) could accept supports the majority's doing so. A point I make in the next chapter is relevant here: It remains obscure why we do not show others the respect that is their due as human beings—or, at least, as "free and equal citizens"—when we offer them, in explanation, what we take to be our true and best reasons for acting as we do (so long as our reasons do not themselves assert, imply, or presuppose the inferior humanity of those to whom the explanation is offered).[67] "There is a gap between a premise which requires the state to show equal concern and respect for all its citizens and a conclusion which rules out as legitimate grounds for coercion the fact that a majority believes that conduct is immoral, wicked, or wrong. That gap has yet to be closed."[68] Merely invoking "the liberal principle of legitimacy" as if it were an axiom of American political morality does not advance the discussion; it does not close the gap. Invoking the principle without defending it will work only for those who already accept the principle. Invoking the principle without defending it does not tell anyone who does not already accept the principle why she should accept it; it does not give anyone reasons, public or otherwise, for accepting it.

The point is not that if a citizen can explain a political choice she wants to make on the basis of premises she believes other citizens could accept, or could reasonably accept, she should not do so. To the contrary, she *should* do so. We have reasons—both a strategic reason and a moral one—for *that* position, as I indicated a few paragraphs back. What we do not yet have are reasons—what Rawls has not given us is an argument—for a different position: the position that if no premises that a citizen believes other (free and equal) citizens could reasonably accept, (and that she is prepared to defend to them as premises they could reasonably accept,) do not support a political choice she wants to make, she should abandon the choice. We need a (non-question-begging) argument for the position that "when basic matters are at stake," a citizen should "appeal only to a public conception of justice and not to the whole truth as they see it". Surely the presumption must be (until Rawls or someone else develops an argument rebutting the presumption) that even "the most fundamental questions should be settled by appealing to the most important truths, [which] may far transcend public reason".[69] Commenting on Rawls's ideal of public reason, John Langan has written: "[I]t is . . . important for [religious groups in a pluralistic society] to retain a certain transcendence in relation to any specific constitutional system, a transcendence which will enable them to protest the atrocities and idolatries of which states have always been capable. To place an idealized form of state over admittedly imperfect communities of faith in our determination of the range of acceptable moral considerations seems to me to be both dangerous and unfaithful, and it is at best a dubious contribution to civic peace."[70] We can all agree, surely, that it is good to cultivate the bonds of political community; it is good to promote what Rawls calls "social unity". But contrary to what Rawls seems to presuppose, social unity is not all-or-nothing; it is more-or-less. The serious inquiry, as Langan's comment suggests, is this: How much social unity—and at what cost or costs? What if, in the view of some citizens—for example, those on the "pro-life" side in the abortion controversy—the lives of innocent human beings hang in the balance? Does Rawls really believe that such citizens should prize an incremental addition to social unity over innocent human life?[71] Instead of Rawls's seeming either/or approach, why not a more nuanced, middle position, according to which, in the interest of promoting social unity, of cultivating rather than subverting the bonds of political community, we try to justify political choices we want to make, as much as possible, on the basis of political "ideals, principles and values that we may reasonably suppose all citizens could accept"—but according to which we do not *invariably* let the inability of a political choice to be justified on such a basis preclude us from making the choice, or from publicly supporting it, on the basis of what we take to be our best reasons, even if, alas, they are what Rawls terms "nonpublic"?

Moreover, given the intractable problem of underdeterminacy mentioned earlier, following the path of public reason does not always lead to one and only one position on a contested issue. With respect to many contested issues, it will be necessary, after following the path of public reason to the end, to go on from there on the basis of one or more nonpublic reasons, *whether religious or nonreligious* (*secular*). Rawls believes that the abortion controversy—the question of what public policy regarding abortion ought to be—should be resolved according to the ideal of public reason.[72] (That the abortion controversy presents the sort of fundamental political issue Rawls means to be resolved according to the ideal of public reason is obvious. At the heart of the abortion controversy, after all, is a fundamental—perhaps the fundamental—political issue: Who is a subject of justice?[73]) But the ideal of public reason is conspicuously underdeterminate with respect to the abortion controversy. This is reflected by the fact that many persons on *both* sides of the controversy—many persons who are "pro-life" *and* many persons who are "pro-choice"—affirm these two relevant, fundamental "public" values: the great worth of human life and the full and therefore equal humanity of women. With respect to the abortion controversy, it is necessary, after following the ideal of public reason to the end, to go on from there on the basis of one or more nonpublic reasons. Some citizens believe that *all* human life, *including unborn human life*, is sacred. What would Rawls have the citizen do who sincerely believes that innocent lives hang in the balance? It bears emphasis here that following the path of public reason does not lead, without the intervention of nonpublic reasons, to the "pro-choice" position in the abortion controversy any more than it leads to the "pro-life" position.

The path of public reason runs out before the "pro-choice" position is reached. Reliance partly on a nonpublic reason or reasons, whether religious or secular, is necessary for those on the "pro-choice" side of the debate no less than for those on the "pro-life" side: for example, "A human fetus is not a 'person' and therefore does not have the rights that persons have."[74]

Because of the role that religious arguments about the morality of human conduct inevitably play in the political process, it is important that such arguments, no less than secular moral arguments, be presented in—so that they can be tested in—public political debate. Moreover, it is impossible to construct "an airtight barrier" between, on the one side, public culture generally—in which religiously based moral discourse is undeniably proper—and, on the other, public debate specifically about controversial political issues. Because Kent Greenawalt and John Rawls have each defended a position less congenial to the airing of religious arguments in public political debate than the position—the *inclusivist* position—I have defended in this chapter, I have detailed here what I take to be the inadequacies both of the arguments Greenawalt gives for his position and of those Rawls gives for his.

Chapter Three

RELIGIOUS ARGUMENTS AS A BASIS OF POLITICAL CHOICE

In this chapter, I continue to pursue the question of the morally proper role of religious arguments about morality in politics. As I said at the beginning of the preceding chapter, by "morally" proper role, I mean simply the role that, taking into account every relevant consideration (other than constitutionality), we should deem it permissible or proper for religious arguments to play in politics. Whereas in the preceding chapter the "we" referred principally to those religious nonbelievers who are skeptical about welcoming religiously based moral discourse into the political discourse of a liberal democracy like the United States, in this chapter the "we" refers principally (though not exclusively) to those religious believers—in particular, to those Christians—who are skeptical about accepting any limits on religiously based moral arguments as a basis of political choice. In speaking to such believers, I address them not from the outside, as a religious nonbeliever speaking to religious believers. Rather, I address them as a religious believer; as I said in the preface to this book, I speak to them (and to others) as a Christian.

According to the construal of the nonestablishment norm I defended in chapter 1, government may rely on a religious argument in making a political choice about the morality of human conduct only if a plausible secular rationale supports the choice. The question remains, however, whether, all things considered, it isn't morally inappropriate for legislators and other public officials, and for citizens voting in a referendum or an initiative election, to rely on a religious argument in making a political choice about the morality of human conduct even if a plausible secular rationale—or even, in their view, a persuasive secular rationale—supports the choice. (I have suggested that those with the principal policymaking authority and responsibility—in particular, legislators—should ask themselves whether they find a secular rationale persuasive.[1]) From the other side, the question remains whether, *apart from the nonestablishment norm*, it isn't morally permissible for legislators and others to rely on a religious argument in making a political choice about the morality of human conduct even if, in their view, no persuasive or even plausible secular rationale supports the choice.

I. Religious Arguments as a Basis of Political Choice

Why might one be inclined to conclude that government should not rely on religious arguments in making political choices about the morality of human conduct? Two reasons come to mind, one of which is moral in character, the other of which is practical. (Although the moral reason might be directed specifically at religiously based moral arguments, it is typically directed at moral arguments without regard to whether they are religious or secular.) According to the moral reason, for government, in making a political choice (or, at least, a coercive political choice), to rely on a moral argument that some persons subject to the choice reasonably reject is for government to deny to those persons the respect that is their due as human beings (or, as Rawls puts it, as "free and equal" persons[2]). This position is deeply problematic. The following comment by William Galston, though it somewhat misconceives the position, goes to the heart of the matter:

> [Charles] Larmore (and Ronald Dworkin before him) may well be right that the norm of equal respect for persons is close to the core of contemporary liberalism. But while the (general) concept of equal respect may be relatively uncontroversial, the (specific) conception surely is not. To treat an individual as person rather than object is to offer him an explanation. Fine; but *what kind* of explanation? Larmore seems to suggest that a properly respectful explanation must appeal to beliefs already held by one's interlocutors; hence the need for neutral dialogue. This seems arbitrary and implausible. I would suggest, rather, that we show others respect when we offer them, as explanation, what we take to be our best reasons for acting as we do.[3]

Let me offer two friendly amendments to Galston's comment. First, it is never to show respect for a human being for one person to offer to another—for example, for a Nazi to offer to a Jew—a reason to the effect that "You are not truly or fully human", even if the Nazi sincerely takes that to be his best reason for acting as he does. Second, Larmore's position, which

Galston somewhat misconceives, is that political "justification must appeal, not simply to the beliefs that the other happens to have, but to the beliefs he has on the assumption (perhaps counterfactual) that he affirms the norm of equal respect."[4] Nonetheless, it remains altogether obscure why we do not give to others the respect that is their due as human beings "when we offer them, as explanation, what we take to be our best reasons for acting as we do" (so long as our reasons do not assert, presuppose, or entail the inferior humanity of those to whom the explanation is offered). According to Robert Audi, "If you are fully rational and I cannot convince you of my view by arguments framed in the concepts we share as rational beings, then even if mine is the majority view I should not coerce you."[5] But *why*? As Gerald Dworkin has observed, "There is a gap between a premise which requires the state to show equal concern and respect for all its citizens and a conclusion which rules out as legitimate grounds for coercion the fact that a majority believes that conduct is immoral, wicked, or wrong. That gap has yet to be closed."[6]

According to a second, practical reason for wanting government to forgo reliance on religiously based moral arguments, the social costs of government relying on such arguments in making political choices (or, at least, coercive political choices)—costs mainly in the form of increased social instability—are too high. It is implausible to believe that in the context of a liberal democratic society like the United States, governmental reliance on religiously based moral arguments in making political choices (even coercive ones) is *invariably* destabilizing—or that it is invariably *more* destabilizing than governmental reliance on controversial secular moral arguments. Some imaginable instances of political reliance on a religiously based moral argument might, with other factors, precipitate social instability. However, "[c]onditions in modern democracies may be so far from the conditions that gave rise to the religious wars of the sixteenth century that we no longer need worry about religious divisiveness as a source of substantial social conflict."[7] John Courtney Murray warned against "project[ing] into the future of the Republic the nightmares, real or fancied, of the past."[8] As Murray's comment suggests, a rapprochement between religion and politics forged in the crucible of a time or a place very different from our own is not necessarily the best arrangement for our time and place. "[W]hat principles of restraint, if any, are appropriate may depend on time and place, on a sense of the present makeup of a society, of its history, and of its likely evolution."[9]

In my view, neither of the two reasons just examined—neither the moral reason nor the practical reason—offers adequate support for the proposition that government should never rely on religious arguments in making political choices about the morality of human conduct. Nonetheless, it does seem to me that in making a political choice about the morality of human conduct, and in the absence of an independent secular rationale for the choice that they find persuasive, legislators and other public officials (and citizens, too) should not rely on—at the very least they should be exceedingly wary about relying on—one sort of religious argument about the morality of human conduct: religious argument about human well-being. Religious argument about human worth, however, is a different matter.

Religious arguments about the morality of human conduct typically address one or both of two fundamental moral issues. First: Are all human beings sacred, or only some; does the well-being of every human being merit our respect and concern, or only the well-being of some human beings? (Even if one accepts that all human beings are sacred, a different but related question can arise: Who is a human being; who is truly and fully human? Women? Nonwhites? Jews?) Second: What are the requirements of human well-being; what is friendly to human well-being, and what is hostile to it; what is good for human beings, and what is bad? There are, correspondingly, two basic kinds of religious argument about the morality of human conduct: religious argument about who among all human beings are sacred and religious argument about the requirements of human well-being. (There can also be religious argument about who is a human being. Consider, for example, the religious argument that God "ensouls" a human fetus at conception and that a human fetus is therefore a human being in the relevant sense. I comment on such an argument later in this chapter.[10]) The claim I want to develop and defend in this chapter is that in making a political choice about the morality of human conduct, neither legislators nor other public officials should rely on a religious argument about the requirements of human well-being unless an independent secular argument reaches the same conclusion about the requirements of human well-being. (The secular argument must be one that, in a legislator's or other public official's *own* view, is persuasive.) I want to turn first, however, to religious arguments about human worth, which, as I said in chapter 1, present a special case.

II. Religious Arguments about Human Worth

The only claim about human worth consistent with the international law of human rights, and the only claim about human worth on which government in the United States constitutionally may rely, is that all human beings, and not just some (e.g., white persons), are sacred.[11] (Let's put aside, for the moment, controversies about who is a human being—in particular, the controversy about whether a human fetus is a human being in the relevant sense.[12]) Claims to the effect that all human beings are sacred are quite common in the United States, where the most influential religious traditions teach that all human beings are children of God and sisters and brothers to one another. (As Hilary Putnam has noted, the moral image central to what Putnam calls the Jerusalem-based religions "stresse[s] equality and also fraternity, as in the metaphor of the whole human race as One Family, of all women and men as sisters and brothers."[13]) The opening passage of a recent statement by the Catholic bishops of Florida, on the controversial political issue of welfare reform, is illustrative: "The founding document of our nation says that all are endowed by their Creator with inalienable rights, including the right to life, liberty, and the pursuit of happiness. And as Jesus has told us: 'Amen, I say to you, whatever you did for the least of these brothers and sisters of mine you did for me.'"[14]

Moreover, claims that all human beings are sacred are quite common not just in the United States, but throughout the world. Indeed, the first part of the idea of human rights—an idea that has emerged in international law since the end of World War II and that is embraced by many persons throughout the world who are not religious believers as well as by many who are—is that each and every human being is sacred. (The second part of the idea is that, because every human being is sacred, there are certain things that ought not to be done to any human being and certain other things that ought to be done for every human being.)[15] "The International Bill of Human Rights", as it is sometimes called, consists of three documents. The first of these, the Universal Declaration of Human Rights (1948), speaks, in the Preamble, of "the inherent dignity . . . of all members of the human family" and of "the dignity and worth of the human person".[16] In Article I, the Declaration proclaims: "All human beings . . . should act towards one another in a spirit of brotherhood." The second and third documents are the International Covenant on Civil and Political Rights (1976) and the International Covenant on Economic, Social and Cultural Rights (1976). The Preamble common to both covenants echoes the Universal Declaration in speaking of "the inherent dignity . . . of all members of the human family". The Preamble then states: "[T]hese rights derive from the inherent dignity of the human person. . . ." The Vienna Declaration and Programme of Action, adopted on June 25, 1993, by the UN-sponsored World Conference on Human Rights,[17] reaffirms this language in insisting that "all human rights derive from the dignity and worth inherent in the human person . . .". A regional human rights document, the American Declaration of the Rights and Duties of Man (1948), begins: "The American peoples have acknowledged the dignity of the individual. . . . The American states have on repeated occasions recognized that the essential rights of man are not derived from the fact that he is a national of a certain state, but are based upon attributes of his human personality. . . ." The Preamble to the American Declaration proclaims: "All men . . . should conduct themselves as brothers to one another." Another regional document, the American Convention on Human Rights (1978), echoes the American Declaration in stating, in the Preamble, that "the essential rights of man are not derived from one's being a national of a certain state, but are based upon attributes of the human personality. . . ." Similarly, the African [Banjul] Charter on Human and Peoples' Rights (1986) says, in the Preamble, that "fundamental human rights stem from the attributes of human beings. . . ." That *every* human being is sacred, and not just some human beings, is emphasized in the Universal Declaration of Human Rights (1948) and in many other international human rights documents by this statement: "Everyone is entitled to all the rights and freedoms set forth in this Declaration, without distinction of any kind, such as race, colour, sex, language, religion, political or other opinion, national or social origin, property, birth or other status."[18]

Of course, the proposition that each and every human being is sacred is, for many persons, a religiously based tenet.[19] However, many persons who are not religious believers embrace the proposition as a fundamental principle of morality. The proposition is an axiom of many secular moralities as well as a fundamental principle, in one or another version, of many religious moralities. The widespread secular embrace of the idea of human rights is conclusive evidence of that fact. As Ronald Dworkin has written: "We almost all accept . . . that human life in all its forms is *sacred*. . . . For some of us, this is a matter of religious faith; for others, of secular but deep philosophical belief."[20] Indeed, the proposition that every human being is sacred is axiomatic for so many secular moralities that many secular moral philosophers have come to speak of "the moral point of view" as that view according to which "every person [has] some sort of equal status".[21] Bernard Williams has noted that "it is often thought that no concern is truly moral unless it is marked by this universality. For morality, the ethical constituency is always the same: the universal constituency. An allegiance to a smaller group, the loyalties to family or country, would have to be justified from the outside inward, by an argument that explained how it was a good thing that people should have allegiances that were less than universal."[22]

Recall, from chapter 1, that under the nonestablishment norm, government may not rely on a religious argument in making a political choice about the morality of human conduct unless a plausible secular rationale supports the choice. I have elsewhere called attention to the possibility that no intelligible secular argument supports the claim that every human being is sacred—that the only intelligible arguments in support of the claim are religious in character.[23] (That an argument is intelligible does not mean that it is persuasive or even plausible.) Let us assume, for the sake of argument, that no intelligible, much less persuasive, secular argument supports the claim that every human being is sacred. It would be silly to insist that because no intelligible secular argument supports the claim that every human being is sacred, the nonestablishment norm forbids government, in making a political choice about the morality of human conduct, to rely on a religious argument that every human being is sacred. Similarly, it would be silly to insist that, as a political-moral matter, citizens, legislators, and other public officials should not rely on a religious argument that every human being is sacred. After all, the proposition that every human being is sacred is a fundamental constituent of American moral culture. (The Declaration of Independence states that "all men are created equal, and endowed by their Creator with certain inalienable rights . . .".) Moreover, the proposition is a fundamental part of the Constitution itself, if the Fourteenth Amendment forbids, as arguably it does, government to base any political choice on the proposition that only some human beings—white persons, for example—are sacred.[24] Therefore, we must conclude that government may, under the nonestablishment norm, and that legislators and others may, as a matter of political morality, rely on a religious argument that every human being is sacred *whether or not any intelligible or persuasive or even plausible secular argument supports the claim about the sacredness of every human being.*

The statement about the nonestablishment norm that begins the preceding paragraph must be qualified, then, with this proviso: In making a political choice about the morality of human conduct, government may rely on a religious argument that every human being is sacred even if no plausible secular argument supports the claim about the sacredness of every human being. This qualification should trouble few if any religious nonbelievers, however. Because the proposition that every human being is sacred is shared not only among many different religious traditions[25] but also between many religious believers and many who have no religious beliefs, laws and public policies that are rooted in the view that every human being is sacred—indeed, that are political representations, legal embodiments, of the view that every human being is sacred—do not in and of themselves privilege either one religion (as such) over another or even religion over nonreligion. By contrast,[26] a law requiring persons to say the Lord's Prayer, for example, *would* privilege not merely religion over nonreligion but also one religion over others: The law would privilege a Christian prayer over other prayers; it would "establish" a Christian prayer.

It remains the case, however, that apart from the qualification just noted, the nonestablishment norm forbids government to rely on a religious argument in making a political choice about the morality of human conduct unless a plausible secular rationale supports the choice. Given the qualification, and given that the other basic sort of religious argument about the morality of human conduct is religious argument about the requirements of human well-being, we may say that the nonestablishment norm forbids government to rely on a religious argument about the requirements of human well-being in making a political choice about the morality of human conduct unless a plausible secular rationale supports the choice. (A qualification is needed here. As I explained in chapter 1, the nonestablishment norm forbids government to base political choices on secular arguments of a certain sort, namely, secular arguments to the effect that one or more religious tenets are more authentically American, or more representative of the sentiments of the community, or otherwise better, than one or more competing religious or nonreligious tenets. When I refer, in describing the requirements of the nonestablishment norm, to a "secular" argument or rationale, I do not mean to include arguments of the sort described in the preceding sentence, but only those that do not in any way valorize one or more religious tenets—that do not claim that one or more religious beliefs are better, along one or another dimension of value, than one or more competing religious or nonreligious beliefs.) In the next section of this chapter, I inquire whether, as a matter not of constitutionality but of political morality, legislators and other public officials should rely on a religious argument about the requirements of human well-being in making a political choice about the morality of human conduct if, in their view, no persuasive secular argument reaches the same conclusion about those requirements as the religious argument.

The political controversy about abortion—the debate about what public policy regarding abortion should be—looms large in the background, it looms large as a subtext, of the debate about the proper role of religion in politics. As much as any other contemporary political controversy, and more than most, the abortion controversy is a principal, if often unspoken, occasion of the debate about religion in politics.[27] It is noteworthy that were government to choose to outlaw abortion, it would not have to rely on a religious argument about the requirements of human well-being. (Therefore, the nonestablishment norm, at least, does not stand in the way of restrictive abortion legislation.[28]) This is illustrated by the fact that the most influential religious voice in the United States on the "pro-life" side of the debate about what public policy regarding abortion should be—the voice of the National Conference of Catholic Bishops—does not rely, at the crucial and controversial stage of its case, on a religious argument about the requirements of human well-being. The bishops' argument comprises three steps:

- Because each and every human being is sacred, the intentional killing of any human being—or, at least, of any innocent human being—is morally forbidden.[29]
- Because there is no nonarbitrary way to draw the bounds of the human community at any point short of conception, we must treat a human fetus as a member of the human community—as a human being, that is, albeit a human being at an early stage of development.
- The intentional killing of any human fetus, therefore, is morally forbidden.[30]

The first step of the bishops' case relies on a claim, which for the bishops is, of course, a religiously grounded claim, that every human being is sacred. This is the same claim the bishops rely on in speaking out about what they perceive to be any human rights abuse. The bishops also rely, in their first step, on a claim about the requirements of human well-being—namely, that killing someone is antithetical to the victim's well-being—but *that* claim about human well-being is religious neither in its content nor in its grounding. The crucial second step of the bishops' case, although controversial in the United States, even among religious believers,[31] gives even many on the "pro-choice" side of the abortion debate pause. As one of the most prominent "pro-choice" theorists, Laurence Tribe, has written, "[T]he fetus is alive. It belongs to the human species. It elicits sympathy and even love, in part because it is so dependent and helpless."[32] In any event, the second step does not involve any argument, religious or secular, about the requirements of human well-being. Nor does it involve a religious argument that the human fetus is a human being in the relevant sense—for example, the argument that God has "ensouled" the human fetus, and that therefore the human fetus should be treated with the same respect and concern with which born human beings should be treated.[33] The second step of the bishops' argument simply suggests what even a religious nonbeliever might want to suggest, namely, that there is no nonarbitrary way to draw the bounds of the human community at any point short of conception. That point—that argument—is not religious. It is not a point that presupposes even that God exists, much less anything about the nature, will, or activity of God. In representing the bishops' position, with which he agrees, Robert George has written: "Opponents of abortion . . . view all human beings, including the unborn . . . , as members of the community of subjects to whom duties in justice are owed. . . . The real issue of principle between supporters of abortion

. . . and opponents . . . has to do with the question of who are subjects of justice." In George's view, "The challenge to the orthodox liberal view of abortion . . . is to identify nonarbitrary grounds for holding that the unborn . . . do not qualify as subjects of justice."[34]

It sometimes seems that some of those who want to marginalize the role of religion in politics hope that by doing so they can gain an advantage in the political debate about abortion. It must be sobering for such persons to learn that the most influential religious voice in the United States on the pro-life side of the abortion debate does not even press, in public political debate, a sectarian religious argument. If the Catholic bishops need not and do not rely on a religious argument about the requirements of human well-being, then government, were it to enact a pro-life position into law, would not need to rely on any such argument.[35]

III. Religious Arguments about Human Well-Being

Arguments about the requirements of human well-being are the second basic kind of religious argument about the morality of human conduct: arguments about what must not be done to, or what must be done for, a human being (including oneself) if she is to flourish—if she is to achieve the greatest well-being of which she is capable—as a human being; what is friendly to authentic human flourishing, or what is hostile to it; what is truly good for human beings, whether all human beings or merely some, or what is truly bad for them. (Not that we can't imagine other sorts of religious arguments about the morality of human conduct. We can. In the context of the abortion controversy, a third sort comes readily to mind: arguments about who is a human being.[36]) Again, the nonestablishment norm forbids government to rely on a religious argument about the requirements of human well-being in making a political choice about the morality of human conduct unless a plausible secular rationale supports the choice. But, apart from the nonestablishment norm, should legislators and other public officials refrain from relying on a religious argument about the requirements of human well-being absent a persuasive (to them) secular argument that reaches the same conclusion about those requirements?

As I am about to explain, for most religious believers in the United States, at least, and probably for most religious believers in other advanced industrial democracies that are religiously pluralistic, the persuasiveness or soundness of any religious argument about the requirements of human well-being depends, or should depend, partly on there being at least one persuasive secular argument (i.e., one secular argument that *they themselves* find persuasive) that reaches the same conclusion about the requirements of human well-being as the religious argument. (Some theologically conservative Christians—in particular, "fundamentalist" Christians and some "evangelical" Christians—will disagree. I address them in the concluding section of this chapter.) A qualification is necessary here. Imagine a religious argument according to which human well-being requires, among other things, prayer or other spiritual practice conducive to achieving knowledge of or union with God. By definition, no "secular" argument can reach such a conclusion about the requirements of human well-being.[37] But, as I explained in chapter 1, no government committed to the ideal of nonestablishment will take any action based on the view that a practice or practices are, as religious practice—practice embedded in and expressive of one or more religious beliefs—truer or more efficacious spiritually or otherwise better than one or more other religious or nonreligious practices or than no religious practice at all. Nonetheless, to be as precise as possible, I should say: The persuasiveness of any religious argument about the requirements of human well-being—any religious argument, that is, on which a government committed *not* to discriminate in favor of religious practice would be prepared to rely—should depend in part on there being at least one sound secular argument that reaches the same conclusion as the religious argument. At least, no religious argument about the requirements of human well-being should be deemed sufficiently strong to ground a political choice, least of all a coercive political choice, unless a persuasive secular argument reaches the same conclusion about the requirements of human well-being.

Why should the persuasiveness of every religious argument about the requirements of human well-being (on which a government committed not to discriminate in favor of religious practice would be prepared to rely) depend partly on there being at least one sound secular route to the religious argument's conclusion about the requirements of human well-being? A "religious" argument about the requirements of human well-being—like a religious argument about anything—is, as I indicated in chapter 1, an argument that relies on (among the other things it relies on) a religious belief: an argument that presupposes the truth of a religious belief and includes that belief as one of its essential premises. (As I said in chapter 1, a "religious" belief is, for present purposes, either the belief that God exists—"God" in the sense of a transcendent reality that is the source, the ground, and the end of everything else—or a belief about the nature, the activity, or the will of God.) The paradigmatic religious argument about the requirements of human well-being relies (partly) on a claim about what God has revealed. Such an argument might be made by someone who believes that we human beings are too fallen (too broken, too corrupt) to achieve much insight into our own nature and that the safest inferences about human nature, about the requirements of human well-being, are based on God's revelation.[38] However, religious believers—even religious believers within the same religious tradition—do not always agree with one another about what God has revealed. Moveover, many religious believers understand that human beings are quite capable not only of making honest mistakes, but even of deceiving themselves, about what God has revealed—including what God might have revealed about the rquirements of human well-being.

(Charles Curran, the Catholic moral theologian, has raised a helpful question, in correspondence, about my "emphasis on human well-being and human nature. Some people might criticize that [emphasis] as being too anthropocentric and not theocentric enough for a truly Protestant position. . . . The primary question perhaps even in the reformed tradition is what is the will of God and not what is human flourishing or human nature."[39] But given two assumptions that few Christians would want to deny, the distinction between doing "what God wills us to do" and doing "what is conducive to the fulfillment of our nature" is quite false. The two assumptions are, first, that human beings have a nature—indeed, a nature fashioned by God—and, second, that it is God's will that human beings act so as to fulfill or perfect their nature.[40])

Therefore, and as many religious believers understand, an argument about the requirements of human well-being—about what is truly good for (all or some) human beings, or about what is truly bad for them—that is grounded on a claim about what has revealed is highly suspect if there is no secular route to the religious argument's conclusion about the requirements of human well-being. So long as there is no persuasive secular argument that supports the conclusion about the requirements of human well-being reached by a religious argument of the kind in question, the religious argument is problematic.[41] Indeed, so long as there is no persuasive secular argument, the religious argument is of doubtful soundness for anyone who believes, as do most Christians, for example, that no fundamental truth about the basic requirements of human well-being is unavailable to religious nonbelievers—that every such truth, even if available only to some human beings by the grace of "supernatural" revelation, is nonetheless available "in principle" to every human being, including nonbelievers, by virtue of so-called "natural" reason.[42] The Roman Catholic religious-moral tradition has long embraced that position:[43]

> Aquinas remained . . . convinced that morality is essentially rational conduct, and as such it must be accessible, at least in principle, to human reason and wisdom. . . . In the teaching of Aquinas, the purpose of revelation, so far as morality is concerned, appears to be essen-

> tially remedial, not absolutely necessary for man . . . [T]he Christian revelation contains in its moral teaching no substantial element over and above what is accessible to human reason without revelation. . . . Revelation as such has nothing in matters of moral behaviour to add to the best of human thinking. . . .[44]

Of course, Aquinas's enormous influence on the Christian religious-moral tradition extends far beyond just Catholic Christianity. Christians generally, and not just Catholics, would "want to argue (at least, many of them would) that the Christian revelation does not require us to interpret the nature of man in ways for which there is otherwise no warrant but rather affords a deeper understanding of man as he essentially is."[45] Moreover, as the American philosopher Robert Audi (who identifies himself as a Christian) has explained, "good secular arguments for moral principles may be *better* reasons to believe those principles divinely enjoined than theological arguments for the principles, based on scripture or tradition." This is because the latter—in particular, scripture-based and tradition-based religious arguments—are "more subject than the former to extraneous cultural influences, more vulnerable to misinterpretation of texts or their sheer corruption across time and translation, and more liable to bias stemming from political or other nonreligious aims." (Christianity's acceptance of slavery comes to mind here—an acceptance that persisted for most of the two millennia of Christianity.[46]) Audi's conclusion: "[I]t may be better to try to understand God through ethics than through theology."[47]

No religious argument about the requirements of human well-being is a persuasive basis of political choice for religious nonbelievers. But even for religious believers—in particular, for religious believers, whether Christian or not, who accept what has been the dominant Christian understanding of the relation between "revelation" and "reason"—any religious argument about the requirements of human well-being should be a highly suspect basis of political choice if no persuasive secular argument reaches the same conclusion about the requirements of human well-being as the religious argument. Given the demonstrated, ubiquitious human propensity to be mistaken and even to deceive oneself about what God has revealed, the absence of a persuasive secular argument in support of a claim about the requirements of human well-being fairly supports a presumption that the claim is probably false, that it is probably the defective yield of that demonstrated human propensity. At least, it fairly supports a presumption that the claim is an inappropriate ground of political choice, especially coercive political choice.

Of course, a religious community might try to insulate itself from such a presumption by means of doctrines about its own privileged and perhaps even infallible insight into God's revelation, including God's revelation about the requirements of human well-being. But such doctrines, which cannot be politically effective in a society as religiously pluralistic as the United States, are destined to seem to outsiders to the community—and, depending on the degree of historical self-awareness among the members of the community, even to some, and perhaps to many, insiders—as little more than hubristic and self-serving stratagems.[48] Moreover, no religious community that fails to honor the ideal of self-critical rationality can play a meaningful role in the politics of a religiously pluralistic democracy like the United States.[49] As Richard John Neuhaus has warned: "So long as Christian teaching claims to be a privileged form of discourse that is exempt from the scrutiny of critical reason, it will understandably be denied a place in discussions that are authentically public."[50] Insisting on a persuasive secular argument in support of a claim about the requirements of human well-being is obviously one important way for the members of a religious community to honor the ideal of self-critical rationality. It is also one important way—and, indeed, a relatively ecumenical way—for the citizens of a religiously pluralistic democracy to test the various statements about what God has revealed, including statements about God's revealed will, that are sometimes articulated in public political debate.

I have just indicated why, in making political choices about the morality of human conduct, legislators and other public officials should not rely on—at least, they should be exceedingly wary about relying on—a religious argument about human well-being if, in their view, no persuasive secular argument reaches the same conclusion about the requirements of human well-being. Should we go further and conclude that legislators and others should not rely on a religious argument about human well-being even if, in their view, a persuasive secular argument *does* reach the same conclusion? Should legislators and others rely *only* on the persuasive secular argument? Recall here the point I emphasized in chapter 1, in responding to the question "Why nonestablishment?": History teaches us to be deeply skeptical about government (about politics, about the politically powerful) acting as an arbiter of *religious* truth. History teaches us to be skeptical as well about government acting as an arbiter of *moral* truth, but there is no way that even a government of very limited powers can avoid making some moral judgments. By contrast, there is simply no need for government to make religious judgments about the requirements of human well-being. As I said in chapter 1, politics is not a domain conducive to the discernment of theological truth; it is, however, a domain extremely vulnerable to the manipulative exploitation of theological controversy. (Theologically conservative Christians, whom I address in the concluding section of this chapter, should know this as well as anyone else.) Nonetheless, it seems unrealistic to insist that legislators and others support a political choice about the morality of human well-being only on the basis of a secular argument they find persuasive if they also find persuasive a religious argument that supports the choice. How could such a legislator be sure that she was relying *only* on the secular argument, putting no weight whatsoever on the religious argument? She could ask whether she would support the choice even if the religious argument were absent, solely on the basis of the secular argument. However, trying to ferret out the truth by means of such counterfactual speculation is perilous at best and would probably be, as often as not, self-deceiving and self-serving.

More fundamentally, the relevant question for a legislator (or other policymaker or citizen) is *not* whether she would find persuasive a secular argument about the requirements of human well-being if she did not already find persuasive a religious argument that reaches the same conclusion as the secular argument. To ask herself that question would be for the legislator to ask herself whether she would find the secular argument persuasive if she were someone other than the person she is, someone without the particular religious beliefs she has. Such a counterfactual inquiry is not only often hopelessly difficult, but, more importantly, beside the point: The proper question is not whether *someone else* would find the secular argument persuasive, but whether, on reflection, *she* finds it persuasive. (I amplify this point, which bears emphasis, in the concluding section of this chapter.) The question she should ask herself is whether, in addition to the religious argument she accepts, she finds persuasive a secular argument that reaches the same conclusion about the requirements of human well-being.[51]

Further considerations buttress the conclusion that no legislator or other public official should rely on a religious argument about human well-being in making a political choice about the morality of human conduct absent a persuasive secular argument that reaches the same conclusion. In a democratic political community (1) that values public deliberation and the political *communitas* (or, as Rawls puts it, "social unity"[52]) that such deliberation helps to nurture and (2) that is religiously pluralistic, legislators and other public officials should not rely on a religious argument about human well-being unless a persuasive secular argument reaches the same conclusion. As the Dutch theologian Edward Schillebeeckx, who is Catholic, has written: "Even when their fundamental inspiration comes from a religious belief in God, ethical norms . . . must be rationally grounded. None of the participants in [religiously based moral discourse] can hide behind an 'I can see what you don't see' and then require [the] others to accept this norm straight out."[53] Even if we assume for the sake of argument that Schillebeeckx's principle should not govern moral discourse in *all* contexts—for example, in the context of a small, monistic, charismatic religious community—the principle should certainly govern moral discourse in *some* contexts, especially in the context of a large, pluralistic, democratic political community like the United States. In words of J. Bryan Hehir, who, as the principal drafter of the U.S. Catholic bishops' 1983 letter on nuclear deterrence,[54] has some experience in the matter:

> [R]eligiously based insights, values and arguments at some point must be rendered persuasive to the wider civil public. There is legitimacy to proposing a sectarian argument within the confines of a religious community, but it does violence to the fabric of pluralism to expect acceptance of such an argument in the wider public arena. When a religious moral claim will affect the wider public, it should be proposed in a fashion which that public can evaluate, accept or reject on its own terms. The [point] . . . is not to banish religious insight and argument from public life[, but only to] establish[] a test for the religious communities to meet: to probe our commitments deeply and broadly enough that we can translate their best insights to others.[55]

The drafters of *The Williamsburg Charter*, a group that included many prominent religious believers, have articulated a similar contention: "Arguments for public policy should be more than private convictions shouted out loud. For persuasion to be principled, private convictions should be translated into publicly accessible claims. Such public claims should be made publicly accessible . . . because they must engage those who do not share the same private convictions . . . ."[56] Neuhaus, who was instrumental in the drafting of *The Williamsburg Charter*, has cautioned that "publicly assertive religious forces will have to learn that the remedy for the naked public square is not naked religion in public. They will have to develop a mediating language by which ultimate truths can be related to the penultimate and prepenultimate questions of political and legal contest."[57]

Consider what we may call the "ecumenical" function of the practice I am recommending here. For citizens and, especially, their elected representatives to decline to make a political choice about the morality of human conduct unless a persuasive secular argument supports the choice—and, concomitantly, for them to rely at least partly on a secular argument in public political debate about whether to make the choice—helps American politics to maintain a relatively ecumenical character rather than a sectarian one. Such a practice deemphasizes one of the most fundamental things that divides us—religion—and in that sense and to that extent is one way of cultivating, rather than fraying, the bonds of political community. (I have discussed the nature of political community, understood as a "community of judgment", elsewhere—and I have explained why political community, thus understood, is a good.[58] It is difficult to understand why any religious community that honors the ideal of self-critical rationality (as any religious community should) would object to such a practice, given that, as I said, insisting on a persuasive secular argument in support of a claim about the requirements of human well-being is one important way for a religious community to honor the ideal. It is especially difficult to understand why any religious community that values ecumenical dialogue with those outside the community would object to such a practice, which can only serve to facilitate such dialogue.[59] Only a historically naive religious (or other) tradition would doubt the value of ecumenical dialogue, which is, among other things, a profoundly important project for anyone committed to the ideal of self-critical rationality. "There is, of course, much to gain by sharpening our understanding in dialogue with those who share a common heritage and common experience with us. . . . Critical understanding of the [religious] tradition and a critical awareness of our own relationship to it, however, is sharpened by contact with those who differ from us. Indeed, for these purposes, the less they are like us, the better."[60]

For the sake of clarity, let me restate the basic position I am defending here: In making a political choice about the morality of human conduct, especially a coercive political choice, neither legislators nor other public officials nor even citizens should rely on a religious argument about the requirements of human well-being unless, in their view, a persuasive secular argument reaches the same conclusion about the requirements of human well-being as the religious argument. But what about religious argument of a third sort: religious argument about who is a human being—about who is truly, fully human? Consider, in particular, a religious argument to the effect that the human fetus is a human being in the relevant sense—for example, the argument that God has "ensouled" the human fetus, and that therefore the human fetus should be treated with the same respect and concern with which born human beings should be treated? Such an argument is not the argument that all human beings are sacred, nor is the argument about the requirements of human well-being. How should such an argument be treated under the regime I am recommending here? As a practical matter, the question seems unimportant. Even if political reliance on such an argument should be constrained in the way that political reliance on a religious argument about the requirements of human well-being should (in my view) be constrained, it is nonetheless the case that anyone who accepts such an argument—a religious argument that the fetus is a human being in the relevant sense—will almost certainly also accept, and can also always rely on, an additional argument that is not problematic under the regime I am recommending here: the argument (which, as I indicated in the preceding section of this chapter, is not religious) that there is no nonarbitrary way to draw the bounds of the human community short of conception.[61]

Someone might conclude that according to the position I am defending here—the position that political reliance on religious arguments about the requirements of human well-being should be constrained in a certain way—the moral insight achieved over time by the various religious traditions, by the various historically extended religious communities, has at most only a marginal place in public political debate about the requirements of human well-being. (In October 1995, in a homily delivered at a mass in Baltimore, Pope John Paul II asked: "Can the biblical wisdom which played such a formative part in the very founding of your country be excluded from the [political] debate [about the morality of human conduct]? Would not doing so mean that America's founding documents no longer have any defining content, but are only the formal dressing of changing opinion? Would not doing so mean that tens of millions of Americans could no longer offer the contribution of their deepest convictions to the formation of public policy?"[62]) Such a conclusion would be mistaken—for three reasons. First, as I emphasized in chapter 2, there are good reasons not merely for tolerating but for encouraging the airing—and testing—of religiously based moral arguments in public political debate. Second, unlike religious arguments about the requirements of human well-being, religious arguments about human worth—in particular, religious arguments that each and every human being is sacred—are not covered by the position I am defending here. Third, and most important for present purposes, the moral insight, the insight into the requirements of human well-being, achieved over time by a religious tradition, *as the yield of the lived experience of an historically extended human community*, might well have a resonance and indeed an authority that extends far beyond just those who accept the tradition's religious claims. Put another way, many of the most basic claims about the requirements of human well-being made by one or another religious tradition are often made, and in any event can be made, without invoking any religious claim (i.e., any claim about the existence, nature, activity, or will of God). What Catholic moral theologian James Burtchaell has explained about the nature of moral inquiry or discernment in the Catholic religious tradition is true of any religious tradition—though, of course, not every religious tradition will accept it as true:

> The Catholic tradition embraces a long effort to uncover the truth about human behavior and experience. Our judgments of good and evil

> focus on whether a certain course of action will make a human being grow and mature and flourish, or whether it will make a person withered, estranged and indifferent. In making our evaluations, we have little to draw on except our own and our forebears' experience, and whatever wisdom we can wring from our debate with others. . . . Nothing is specifically Christian about this method of making judgments about human experience. That is why it is strange to call any of our moral convictions "religious," let alone sectarian, since they arise from a dialogue that ranges through so many communities and draws from so many sources.[63]

Many religious believers and nonbelievers alike have failed to see the overwhelming extent to which both the development of insight into the requirements of human well-being and the debate that attends such development is, inside religious traditions as much as outside them, nonrevelational and even nontheological. Because the moral insight achieved over time by the various religious traditions is substantially nonrevelational and even nontheological, bringing that insight to bear in a politics constrained by the ideal of nonestablishment is not the problem some religious believers and nonbelievers imagine it to be. The Jesuit priest and sociologist John Coleman has observed, in a passage that reflects Aquinas's influence: "[M]any elements and aspects of a religious ethic . . . can be presented in public discussion in ways that do not presume assent to them on the specific premises of a faith grounded in revelation. Without being believing Hindus, many Westerners, after all, find in Gandhi's social thought a superior vision of the human than that of ordinary liberal premises."[64] Martin Marty has commented, in much the same spirit, that "religionists who do not invoke the privileged insights of their revelation or magisterium can enhance and qualify rationality with community experience, intuition, attention to symbol, ritual, and narrative."[65]

Indeed, to embrace a religious premise—a biblical premise, for example—about what it means to be human, about how it is good or fitting for human beings to live their lives, and then to rely on the premise in public discourse, is not even *necessarily* to count oneself a participant in the religious tradition that has yielded the premise; it is not even necessarily to count oneself a religious believer. You certainly do not have to be Jewish to recognize that the prophetic vision of the Jewish Bible is profound and compelling, any more than you have to be Catholic or Presbyterian or Baptist or even Christian to recognize that the Gospel vision of what it means to be human is profound and compelling. Gandhi was not a Christian, but he recognized the Gospel vision as profound and compelling. As David Tracy has emphasized: "Some interpret the religious classics not as testimonies to a revelation from Ultimate Reality, . . . but as testimonies to possibility itself. As Ernst Bloch's interpretations of all those daydreams and Utopian and eschatological visions that Westerners have ever dared to dream argue, the religious classics can also become for nonbelieving interpreters testimonies to resistance and hope. As Mircea Eliade's interpretations of the power of the archaic religions show, the historian of religions can help create a new humanism which retrieves forgotten classic religious symbols, rituals, and myths."[66] Tracy continues: "If the work of Bloch and [Walter] Benjamin on the classic texts and symbols of the eschatological religions and the work of Eliade and others on the primal religions were allowed to enter into the contemporary conversation, then the range of possibilities we ordinarily afford ourselves would be exponentially expanded beyond reigning Epicurean, Stoic, and nihilistic visions."[67]

So—and I want to emphasize this—it is simply not true that according to the position I am presenting here, the moral insight achieved over time by the various religious traditions, by the various historically extended religious communities, has at most only a marginal place in public political debate about the morality of human conduct. Such insight, as the comments by Burtchaell, Coleman, and Tracy suggest, may play a central role even in a politics constrained by the ideal of nonestablishment.

> But, the objection may be pressed, can a religious body argue its case in a secular forum (i.e., one that is not already antecedently committed to the religion in question)? Either, it may be said, it will rely on Christian premises, which *ex hypothesi* opponents will not accept; or it will employ purely secular premises, in which case the ensuing law will not be Christian. In neither case will any genuine debate have taken place between Christians and non-Christians. The dichotomy, however, is altogether too neat to be convincing. It presupposes that there is and always must be a complete discontinuity between Christian and secular reasoning. Certainly this can occur—if, for example, the Christian is an extreme fundamentalist and the secular thinker regards individual preferences as the sole basis for morality. . . . But, . . . Christians would presumably want to argue (at least, many of them would) that the Christian revelation does not require us to interpret the nature of man in ways for which there is otherwise no warrant but rather affords a deeper understanding of man as he essentially is. If that is so, there is room for a genuine exchange of ideas.[68]

IV. A Case in Point: Religious Argument about the Morality of Homosexual Sexual Conduct

I now want to illustrate, by reference to the political controversy about the morality of homosexual sexual conduct, my point that legislators and other public officials should be extremely wary about relying on a religious argument about the requirements of human well-being in making a political choice about the morality of human conduct if, in their view, no persuasive secular argument reaches the same conclusion. The political controversy in the United States today about the morality of homosexual sexual conduct—which is at the center of the debate about whether the law should recognize homosexual marriage or at least grant some sort of marriage-like status to same-sex unions[69]—is, like the political controversy about the morality of abortion, a principal context for the debate about the proper role of religion in politics. Moreover, the controversy is at its core about the requirements of human well-being. According to the explicit or implicit position of those on one side of the controversy, the fundamental reason why homosexual sexual conduct is invariably immoral is because such conduct is invariably antithetical to, subversive of, the authentic well-being—the authentic flourishing—of anyone who engages in it. They believe that such conduct is, in that sense, unworthy of anyone who would be truly, fully human.

Just as it is implausible to suggest that all heterosexual sexual conduct is moral, it is implausible to suggest that all homosexual sexual conduct is moral. Homosexual sexual conduct, like heterosexual sexual conduct—even heterosexual sexual conduct between persons married to one another—can be exploitative, abusive, self-destructive, and so on.[70] The serious question is whether some homosexual sexual conduct, like some heterosexual sexual conduct, can be moral, or whether all such conduct is immoral, *even homosexual sexual conduct that is embedded in and expressive of a lifelong, monogamous relationship of faithful love—indeed, that is a generative matrix of such a relationship, of such love.*[71]

Consider the religious argument that God has revealed that all homosexual sexual conduct is immoral. Although many Christians (and other religious believers) accept that argument, a growing number of Christians do not. For example:

> In June 1994, the General Assembly of the Presbyterian Church (U.S.A.) came within a few votes of permitting ministers to bless same-sex unions. Also in June 1994, a draft proposal by the Episcopal bishops, after describing homosexuality as an orientation of "a significant minority of persons" that cannot usually be reversed," went on to say that sexual relationships work best within the context of a committed lifelong union: "We believe this is as true for homosexual relationships as for heterosexual relationships and that such relationships need and should receive the pastoral care of the church." In October 1993, a draft report by a national Lutheran study group on sexuality called for the blessing and even legal acknowledgement of loving gay relationships.[72]

Even in the Catholic Church, a growing number of moral theologians are dissenting from the Church's official position that all homosexual sexual conduct is immoral.[73] Moreover, recent polling data suggests that only about 56 percent of all Catholic priests in the United States accept the Church's official position on the morality of homosexual sexual conduct.[74]

Although many Christians believe that God has revealed, in the Bible, that all homosexual sexual conduct is immoral, many thoughtful Christians reject that interpretation of the Bible.[75] Fundamentalist religious arguments of any kind, including fundamentalist religious arguments against homosexual sexual conduct,[76] are deeply problematic, *even for those who count themselves religious*.[77] To be sure, not every argument against homosexual sexual conduct based on the Bible—whether the Jewish Bible, the New Testament, or both—is a fundamentalist argument. Nonetheless, as an impressive growing literature in contemporary Christian ethics argues, no biblically based argument against homosexual sexual conduct fails to be highly controversial even for many persons who accept the authority of the Bible.[78]

In any event, many Christians (and others) understand that human beings are prone not only to making honest mistakes, but even to deceiving themselves, about what God has revealed. For example, there have been Christian (and other) religious arguments for racist beliefs and for slavery and other racist practices; those arguments have been discredited. There have also been Christian (and other) religious arguments for sexist beliefs and practices; those arguments, too, have been discredited. That today there are Christian (and other) religious arguments for "heterosexist" beliefs and practices[79]—in particular the belief that all homosexual sexual conduct is immoral—does not entail that the arguments are correct or that they will not be discredited; many think that the arguments are already well on the way to being discredited.[80] Because religious believers, like other human beings, are prone both to error and to self-deceit, the religious argument that all homosexual sexual conduct is contrary to what God has revealed in the Bible is highly suspect if there is no secular route to the religious argument's conclusion that all homosexual sexual conduct is immoral. Indeed, if there is no persuasive secular argument in support of that conclusion, the religious argument is highly suspect; it is of doubtful soundness, for anyone who believes—as do Catholics and many other Christians, for example—that fundamental truths about the basic requirements of human well-being are available "in principle" to every human being, including nonbelievers, by virtue of so-called natural reason.

Is there a persuasive secular argument that all homosexual sexual conduct is immoral? John Finnis recently tried to construct a secular argument in support of the traditional religious tenet that all homosexual

sexual conduct is immoral—even homosexual sexual conduct embedded in and expressive of a lifelong, monogamous relationship of faithful love.[81] As I demonstrate in the next section of this chapter, however, Finnis's secular argument is far from persuasive.[82] In the wake of Finnis's failure, one can fairly doubt that any secular argument that all homosexual sexual conduct is immoral is sound. If no such secular argument is persuasive, then, for the reasons I have given in this chapter, no religious argument that all such conduct is immoral should serve as a basis of political choice, least of all as a basis of coercive political choice.

V. Finnis's Secular Argument about the Morality of Homosexual Sexual Conduct

John Finnis, the Professor of Law and Legal Philosophy at Oxford University, is a Roman Catholic. Again, the Roman Catholic tradition has long embraced the conviction that no fundamental truth about the basic requirements of human well-being is unavailable to religious nonbelievers—that every such truth, even if available only to some human beings by the grace of "supernatural" revelation, is nontheless available to every human being, including nonbelievers, by virtue of so-called natural reason. True to that conviction, when Finnis set out to defend the traditional Roman Catholic position that all homosexual sexual conduct is immoral, he sought to provide a secular argument—an argument that, in Finnis's words, is "reflective, critical, publicly intelligible, and rational".[83] In addressing the question "What is wrong with homosexual conduct?", Finnis denies that "the judgment that it is morally wrong [is] inevitably a manifestation either of mere hostility to a hated minority, or of purely religious, theological, and sectarian belief".[84] It is precisely because Finnis's argument aims to be "reflective, critical, publicly intelligible, and rational"—and because Finnis is himself a sophisticated moral philosopher (in the natural law tradition)[85]—that his argument deserves serious examination.[86] If Finnis is unable to provide a persuasive secular argument in support of the position he wants to defend, there is reason to doubt that such an argument exists.[87] As Finnis understands, if no such argument exists, then the Church's traditional position on homosexual sexual conduct—which is also the traditional position of Christianity generally and also of Judaism and of Islam—is in serious doubt.

What is Finnis's secular argument that all homosexual sexual conduct between consenting adults is morally wrong or bad?[88] (At one point Finnis uses the word "evil".[89]) What is his secular argument that all such conduct is immoral "even for anyone unfortunate enough to have innate or quasi-innate homosexual inclinations"?[90] Finnis writes that "[g]enital intercourse between spouses enables them to actualize and experience (and in that sense express) their marriage itself, as a single reality with two blessings (children and mutual affection). Non-marital intercourse, especially but not only homosexual, has no such point and therefore is unacceptable."[91] Gertrude Stein said of Oakland: "There is no 'there' there." We can fairly wonder about Finnis's statement ("Non-marital intercourse . . . therefore is unacceptable") whether there is a "therefore" there. There might be, instead, merely a non sequitur. Why accept Finnis's claim that in order to be "acceptable", sexual intercourse between two persons must enable them to "actualize and experience (and in that sense express)" their relationship "as a single reality with two blessings (children and mutual affection)"? Finnis is right that the relationship of a man and a woman in marriage, a relationship that is partly sexual, is—or, at least, at its best can be—"a single reality". Why insist that no other relationship—that is, no other relationship that is partly sexual—can be, even at its best, a single reality? Why not believe, instead, that any friendship, including one that is partly sexual, can constitute, at its best, a single reality? If any friendship can be, at its best, a single reality, then, depending on the nature of the friendship, which might or might not be partly sexual, one or another act (which, depending on the nature of the friendship, might or might not be sexual) might "enable them to actualize and experience (and in that sense express) their [friendship] itself, as a single reality with [its one or more] blessings", whatever those blessings might be (which depends on the nature of the friendship). Why doubt that at its best, any sexual union between two adults, whether heterosexual or homosexual—that is, a sexual union that is lifelong, monogamous, faithful, and deeply loving—can be a single reality?

Developing his argument, Finnis writes: "[I]n sterile and fertile marriages alike, the communion, companionship, *societas* and *amicitia* of the spouses—their being married—*is* the very good of marriage, and is an intrinsic, basic human good, not merely instrumental to any other good." He then refers to "this communion of married life" as an "integral amalgamation of the lives of two persons. . . ." Finnis concludes this part of his argument by stating, approvingly, "the position that procreation and children are neither the *end* (whether primary or secondary) to which marriage is instrumental (as Augustine taught), nor instrumental to the good of the spouses (as much secular and 'liberal Christian' thought supposes), but rather: Parenthood and children and family are the intrinsic fulfillment of a communion which, because it is not merely instrumental, can exist and fulfill the spouses even if procreation happens to be impossible for them."[92] But why isn't it true of any lifelong, monogamous, faithful, and loving friendship that is partly sexual, whether it be between a man and a woman, a man and a man, or a woman and a woman, that "the communion, companionship, *societas* and *amicitia* of the friends—their being united in their friendship—*is* the very good of their friendship, and is an intrinsic, basic human good, not merely instrumental to any other good"? Why isn't it true of any lifelong, monogamous, faithful, loving friendship that is partly sexual that "this communion of friendship" is an "integral amalgamation of the lives of two persons"?

Consider Finnis's statement that "[p]arenthood and children and family are the intrinsic fulfillment of a communion which, because it is not merely instrumental, can exist and fulfill the spouses even if procreation happens to be impossible for them." No doubt parenthood and children and family are the intrinsic fulfillment of a *certain kind of* communion—one in which the spouses want to have children and raise a family, perhaps even one in which the spouses have joined themselves expressly (though not solely) for that purpose, perhaps because they judge it to be (a part of) their vocation, their calling, to do so. But obviously not every (heterosexual) marriage is a communion *of that kind*; not every marriage is one in which the spouses want to have children; not every pair of spouses judge it to be their vocation to raise a family. Finnis acknowledges that "a communion . . . , because it is not merely instrumental, can exist and fulfill the spouses even if procreation happens to be impossible for them." But why can't a communion exist and fulfill the spouses even if procreation is something they have chosen to forgo? (Perhaps they have chosen to forgo procreation for reasons of health—or perhaps because they have chosen to devote their lives to a demanding ministry that is realistically incompatible with their raising a family of their own. There can be morally *good* reasons, after all, for choosing to forgo procreation—reasons worthy of one who would be truly, fully human—just as there can be morally bad reasons. Even Finnis will allow that, for example, there can be morally good reasons for choosing to forgo procreation and instead become a celibate priest as well as morally bad reasons.) Moreover, why can't a lifelong, monogamous, faithful, and loving communion between a man and a man or a woman and a woman—a communion that is partly sexual—exist and fulfill the partners even though procreation is in the nature of things not available to them (which they might or might not regret, depending on what they would choose if it were available to them)?[93]

Finally, Finnis begins to get to the heart of the matter; he finally articulates the question he must address: "Why cannot non-marital friendship be promoted and expressed by sexual acts? Why is the attempt to express affection by orgasmic non-marital sex the pursuit of an illusion?" Why is it "that homosexual *conduct* (and indeed all extra-marital sexual gratification) is radically incapable of participating in, actualizing, the common good of friendship"?[94] Finnis begins his response to this, the crucial inquiry, with this passage:

> [T]he common good of friends who are not and cannot be married (for example, man and man, man and boy, woman and woman) has nothing to do with their having children by each other, and their reproductive organs cannot make them a biological (and therefore personal) unit. So their sexual acts together cannot do what they may hope and imagine. Because their activation of one or even of each of their reproductive organs cannot be an actualizing and experiencing of the *marital* good—as marital intercourse (intercourse between spouses in a marital way) can, even between spouses who *happen* to be sterile—it can do no more than provide each partner with an individual gratification. For want of a *common good* that could be actualized and experienced *by and in this bodily union*, that conduct involves the partners in treating their bodies as instruments to be used in the service of their consciously experiencing selves; their choice to engage in such conduct thus dis-integrates each of them precisely as acting persons.[95]

The fundamental problem with the foregoing passage is the false claim that even in the context of a homosexual friendship that is a lifelong, monogamous relationship of faithful love, homosexual sexual conduct "can [never] do [any] more than provide each partner with an individual gratification." Homosexual sexual conduct, like heterosexual sexual conduct, *can* do more—much more—than provide each partner with "an individual gratification". Interpersonal sexual conduct, whether heterosexual or homosexual, can be a way of affirming and serving both the sexual and the emotional well-being of one's lover; as such, sexual conduct can both express, in a bodily (embodied) way, one's love for one's lover; indeed, at its best such conduct can be a generative matrix of the emotional strength one needs to live well—to live a truly, fully human life—and therefore to attend to one's most challenging responsibilities, such as those that attend being a parent.[96] Sexual conduct can be all this (and more) even if it is not meant to be—indeed, *even if it is meant not to be*—procreative. Note, moreover, that the (authentic) well-being of one's lover—including both the sexual and the emotional well-being of one's lover—is a *common* good, a good not only for one's lover but also for oneself. This is so because what is (truly) good for one's lover is also good for oneself, just as it is the case that what is good for one's child, what is conducive to or even constitutive of one's child's well-being, is also, *because it is good for one's child*, good for oneself.

Finnis forthrightly acknowledges, in his essay, that his argument is not confined to homosexual sexual conduct, but applies to heterosexual conduct that is, in Finnis's words, "deliberately contracepted".[97] According to Finnis, then, "deliberately contracepted" heterosexual conduct cannot, not even in the context of a marriage, "do . . . more than provide each partner with an individual gratification", and the "choice to engage in such conduct thus dis-integrates each of them precisely as acting persons." One who, on the basis of his or her experience—indeed, perhaps on the basis of his or her experience in marriage—disagrees with Finnis's point as applied to "deliberately contracepted" heterosexual conduct has good reason to be skeptical that Finnis's point as applied to homosexual sexual conduct has any firmer grounding in experience. There is an (inappropriately) abstract quality to Finnis's argument, to the point where he seems to claim an a priori knowledge that it is impossible for either homosexual sexual conduct or "deliberately contracepted" heterosexual sexual conduct (even in the context of a marriage) to do more than provide each partner with an individual gratification. What a strange claim.[98]

In any event, why is the choice of two persons to engage in sexual conduct in mutual satisfaction of their sexual appetite—in particular, the choice of two partners in a lifelong, monogamous relationship of faithful love to engage in sexual conduct in a way that affirms and serves the sexual and emotional well-being of one another—*necessarily* disintegrative of each of them "precisely as acting persons" just in virtue of the fact that the particular sexual conduct they choose to engage in is either nonprocreative or "deliberately contracepted"?[99] Finnis writes:

> The union of the reproductive organs of husband and wife really unites them biologically (and their biological reality is part of, not merely an instrument of, their *personal* reality); reproduction is one function and so, in respect of that function, the spouses are indeed one reality, and their sexual union therefore can *actualize* and allow them to *experience* their *real common good—their marriage* with the two goods, parenthood and friendship, which (leaving aside the order of grace) are the parts of its wholeness as an intelligible common good even if, independently of what the spouses will, their capacity for biological parenthood will not be fulfilled by that act of genital union.[100]

What drives Finnis's position, then, is his view—which is also the view of his mentor-collaborator, Germain Grisez[101]—that any sexual conduct between two persons, even persons married to one another, is morally illicit if it cannot or does not "actualize" and "allow them to experience" their relationship as, at least in part, a *procreative* union (a would-be if not actual procreative union). Let me put this Grisez-Finnis view into perspective for the reader by quoting a remarkable passage from what Finnis describes as "the new second volume of Grisez's great work on moral theology":[102] A married couple's sexual act is morally illicit "if either or both spouses do anything inconsistent with their act's being of itself suited to procreating (for example, *if spouses unable to engage in intercourse due to the husband's impotence masturbate each other to orgasm, if a couple trying to prevent the transmission of disease use a condom, or if either or both spouses do something in order to impede conception*)."[103]

As I said, a remarkable passage. The Grisez–Finnis view is controversial and indeed widely rejected among Christians, even among Catholic Christians, few of whom today deny—indeed, many, probably most, Catholic moral theologians today affirm—that the sexual conduct of a husband and a wife can be morally licit if it "actualizes" and "allows them to experience" their marriage, not, or not *any longer*, or not *yet*, as a procreative union (actual or would-be), but simply, or *now* simply, as a sexual-spiritual union (i.e., a sexually rooted and sexually embodied spiritual union) of profound depth and richness.[104] The nonprocreative sexual conduct of a man and a woman in a lifelong, monogamous relationship of faithful love can be morally licit if it "actualizes" and "allows them to experience" their friendship as a sexual-spiritual union of profound depth and richness.[105] Why, then, can't the sexual conduct of a man and a man or of a woman and a woman also be morally licit—why cannot it also be worthy of those who would be truly, fully human—if it actualizes and allows *them* to experience *their* friendship as a lifelong, monogamous, faithful, loving sexual-spiritual union of profound depth and richness?

The social and legal institutionalization of the position on the morality of homosexual sexual conduct for which Finnis contends has been hostile to the development of a moral culture in which there would be more homosexual relationships that are lifelong, monogamous relationships of faithful love. It would be perverse, therefore, for one who applauds that institutionalization to try to make hay of the fact that many homosexual relationships are not lifelong, monogamous, faithful, and loving. "[A]s Rich Tafel, head of the Log Cabin Republicans, argues in debates with religious-right opponents in his party: 'You can't have it both ways—accusing gays of being promiscuous and then denying us the right to incorporate into monogamous, legally recognized relationships.'"[106] In any event, the question before us is not how many homosexual relationships are lifelong, monogamous relationships of faithful love. Not all heterosexual relationships—not even all heterosexual "marriages"—are such relationships. I do not know how many homosexual relationships are lifelong, monogamous, faithful, and loving, any more than I know how many heterosexual relationships are such relationships. The question is: Regardless of how many or how few homosexual relationships are lifelong, monogamous, faithful, and loving, is homosexual sexual conduct *in the context of such a relationship* necessarily morally illicit?

At this point in his argument, Finnis says something that exemplifies the (inappropriate) abstractness of his analysis. Denying the moral relevance of the sort of considerations—the sort of particularities of context—that are widely agreed, by Christians and others, to be not merely morally relevant, but morally determinative, Finnis writes:

> Reality is known in judgment, not in emotion, and *in reality*, whatever the generous hopes and dreams and thoughts of *giving* with which some same-sex partners may surround their sexual acts, those acts cannot express or do more than is expressed or done if two strangers engage in such activity to give each other pleasure, or a prostitute pleasures a client to give him pleasure in return for money, or (say) a man masturbates to give himself pleasure and a fantasy of more human relationships after a grueling day on the assembly line. . . . [T]here is no important distinction in essential moral worthlessness between solitary masturbation, being sodomized as a prostitute, and being sodomized for the pleasure of it. Sexual acts cannot *in reality* be self-giving unless they are acts by which a man and a woman actualize and experience sexually the real giving of themselves to each other—in biological, affective and volitional union in mutual commitment, both open-ended and exclusive—which like Plato and Aristotle and most peoples we call marriage.[107]

One wonders what Finnis means by "reality" or "in reality". As if the *reality* of sexual acts—the *reality* of what they express and do—could possibly be determined without regard to whether those acts are inspired by, animated by, "generous hopes and dreams and thoughts of *giving*". It is (dare I say it) absurd, even perverse, to suggest that from a moral perspective the reality of sexual conduct that takes place in and is expressive of a lifelong, monogamous homosexual relationship of faithful love—or the reality of "deliberately contracepted" sexual conduct that takes place in and is expressive of a lifelong, monogamous heterosexual relationship of faithful love—is essentially the same as the reality of the sexual conduct of "a prostitute pleasuring a client in return for money". (According to Finnis, "there is no important distinction in essential moral worthlessness" among those different types of sexual conduct.) Responding to the claim that "[h]omosexual acts, by definition and in principle, sever the connection between sexuality and procreation", Thomas Stahel, a Jesuit priest and former executive editor of the Jesuit weekly *America* and now assistant to the president of Georgetown University, writes: "But if homosexual acts are human acts, their significance is not restricted to their physical description or physical effects. If homosexuality, as practiced by Christians, could be shown to be procreative in some way that transcends the biological, we might attempt to assign it a teleology fitted to Christian morality. That is the crux. So it will not do to reduce the significance of human sexuality, whether straight or gay, to the physical act itself."[108]

Finnis's judgment/emotion opposition ("[r]eality is known in judgment, not in emotion") is simplistic and misleading. Sometimes an emotional response can impede one's reaching a sound judgment; sometimes, however, an emotional response not only precipitates the process of judgment, but also clarifies or illuminates the way to a sound judgment.[109] Let us agree, in any event, that "[r]eality is known in judgment". It is nonetheless true that the abstract "reality" that Finnis seems to believe he knows *a priori* is far removed from the reality that many others believe they know *a posteriori*, on the basis of experience, whether their own experience or that of credible others whom they trust. It is far removed from the *experienced* reality that must inform our judgments about what conduct is, or is not, worthy of those who would be fully, truly human. The reality apprehended by many married couples who practice contraception, and by many homosexual couples, is directly contrary to the reality postulated by John Finnis (and by Germain Grisez and the religious-moral tradition Grisez re-presents).

Finnis is reduced to claiming that the reality apprehended by many married couples who practice contraception and by many homosexual couples, unlike the reality asserted by him, is illusory. Finnis refers to the married couple's "illusions of intimacy and self-giving in" acts of deliberately contracepted sex.[110] According to Finnis, the many married couples who engage in "deliberately contracepted" sex, including millions of Christian married couples (many of whom are, like Finnis, Catholic Christians[111]), are, if they think they are not doing something morally wrong, in the grip of an "illusion". In Finnis's view, no doubt, it is an illusion aided and abetted for the Christian couples by all those ministers and priests and theologians who do not submit to the position Finnis defends. By contrast, Finnis is, confidently and happily, in the grip of "reality"—which is known by him not in "emotion", but in "judgment". Only *one-quarter* of all American Catholic priests accepts the Church's official teaching on contraception—and only *a little more than half* of them (56 percent) accepts the Church's position on homosexual sexual conduct.[112] Are all those dissenting priests—many of whom daily minister to married couples or to homosexual couples or to both—in the grip of an "illusion", too?

The reader can decide for herself whose "reality" is an illusion: that apprehended by the heterosexual couples who engage in mutually affirming and nurturing but "deliberately contracepted" sexual conduct in the context of their marriages, and by the homosexual couples who engage in mutually affirming and nurturing sexual conduct in the context of their lifelong, monogamous relationships of faithful love—or, instead, that postulated by John Finnis, for whom "deliberately contracepted" sex in marriage and homosexual sex in a lifelong, monogamous relationship of faithful love is *equal in moral worthlessness* to the commercial sex of a prostitute with the prostitute's client. The serious question, in my view, is not who is in the grip of an "illusion" but why Finnis, unlike so many others, has not been able to break the grip of the particular illusion that holds him, why he has not been able to see through it. Let us recall that in the Grisez–Finnis view of the matter, a married couple's sexual act is "in reality" morally worthless—no less so than is the commercial sex of the prostitute—"if either or both spouses do anything inconsistent with their act's being of itself suited to procreating (for example, *if spouses unable to engage in intercourse due to the husband's impotence masturbate each other to orgasm, if a couple trying to prevent the transmission of disease use a condom, or if either or both spouses do something in order to impede conception*)." That astounding passage, and indeed Finnis's whole argument, radically discount "the authority of Christian experience, for which there can be no substitute."[113] As Margaret Farley has explained, with particular reference to homosexual sexual conduct:

> The final source for Christian ethical insight is [contemporary experience]. . . . I am referring primarily to the testimony of women and men whose sexual preference is for others of the same sex. Here, too, we have as yet no univocal voice putting to rest all of our questions regarding the status of same-sex relations. We do, however, have some clear and profound testimonies to the life-enhancing possibilities of same-sex relations and the integrating possibilities of sexual activity within these relations. We have the witness that homosexuality can be a way of embodying responsible human love and sustaining Christian friendship. Without grounds in scripture, tradition, or any other source of human knowledge for an absolute prohibition of same-sex relations, this witness alone is enough to demand of the Christian community that it reflect anew on the norms for homosexual love.[114]

In the final passages of his argument, Finnis presents two new claims. First, he claims that "[t]he deliberate genital coupling of persons of the same sex . . . is sterile and disposes the participants to an abdication of responsibility for the future of humankind."[115] This is a silly claim. After all, by itself the decision of a homosexual couple living in a lifelong, monogamous relationship of faithful love to engage in sexual conduct with one another no more necessarily disposes them "to an abdication of responsibility for the future of humankind" than by itself their decision not to engage in sexual conduct with one another necessarily disposes them to an acceptance of responsibility for the future of humankind. Indeed, the sexual conduct of a homosexual couple who are raising children,[116] like that of a "deliberately contracepting" heterosexual married couple who are raising children, might well be—and at its best certainly is—a source of the kind of mutual emotional nurture that helps them to raise the children in their charge with great love and strength.[117]

Second, Finnis claims that homosexual sexual conduct, even in the context of a lifelong, monogamous relationship of faithful love,

> treats human sexual capacities in a way which is deeply hostile to the self-understanding of those members of the community who are willing to commit themselves to real marriage in the understanding that its sexual joys are not mere instruments or accompaniments to, or mere compensations for, the accomplishment of marriage's responsibilities, but rather enable the spouses to *actualize and experience* their intelligent commitment to share in those responsibilities, in that genuine self-giving. [It] treats human sexual capacities in a way which is deeply hostile to the self-understanding of those members of the community who are willing to commit themselves to real marriage.[118]

This claim rests on Finnis's false belief that even in the context of a lifelong, monogamous relationship of faithful love, homosexual sexual conduct "can do no more than provide each partner with an individual gratification."[119] Finnis seems oblivious to the fact that the mutually affirming and nurturing sexual love of two homosexual partners for one another—two partners living a lifelong, monogamous relationship of faithful love—does not presuppose (nor does it entail) that "sexual joys [in the context of marriage] are mere instruments or accompaniments to, or mere compensations for, the accomplishment of marriage's responsibilities". The reason that the mutually affirming and nurturing sexual love of two homosexual partners for one another does not presuppose what Finnis imagines it to presuppose is that the partners might well understand, based on their own lived experience, that far from being "mere instruments or accompaniments to, or mere compensations for, the accomplishment of [their relationship's] responsibilities," their sexual joys "enable [them] to *actualize and experience* their intelligent commitment to share in those responsibilities, in that genuine self-giving."

By now it should be very clear that the logic of Finnis's position is such that, according to the position, "deliberately contracepted" sexual conduct by a married couple, no less than homosexual sexual conduct, is "deeply hostile" to the self-understanding to which Finnis refers. But such conduct can scarcely be deeply hostile to the self-understanding of all those very many married couples—including a very substantial majority (80 percent) of Catholic couples of childbearing age in the United States[120]—who regularly engage in what Finnis calls "deliberately contracepted" sexual conduct. The reason that such conduct is not hostile to their self-understanding is because, based on their own lived experience, they understand, though Finnis does not, that "deliberately contracepted" sex *can* do much more "than provide each partner with an individual gratification"—as can homosexual sexual conduct. If, contra Finnis, "deliberately contracepted" sex is not "deeply hostile" to *their* self-understanding, why should homosexual sex that takes place in the context of a lifelong, monogamous relationship of faithful love be deeply hostile to their self-understanding? Compare, to Finnis's claim about the deep hostility of homosexual sexual conduct to the self-understanding to which he refers, this comment by a Catholic "wife and mother" (Joan Sexton) in her letter to the editor of *Commonweal*, the American lay Catholic weekly:

> [Here are some] thoughts about the argument that gay marriage would endanger the institution of heterosexual marriage. . . .
>
> [I]t seems to me, as wife and mother, that it may be the most committed of hearts that would enter and stay in a marriage as a one-to-one relationship. I admire the courage of the homosexual person giving him/her self to one person, one body, one heart and to a lifelong struggle to understand and support that other. It's a promise that draws, from me, at least, respect and awe.
>
> And I suspect that such marriages could teach us a lot in terms of realizing the ideal of true friendship. It's interesting that the challenge of women on ordination has brought forth a fresh new look at what priesthood means; so the challenge of gay couples to be included in the institution of marriage promises a new look at what marriage means.[121]

Joan Sexton seems to have discerned something that has, so far, eluded the grasp of John Finnis.

Recall that Finnis means his argument to be, at every turn, "reflective, critical, publicly intelligible, and rational". (Finnis denies, recall, that "the judgment that [homosexual sexual conduct] is [always] morally wrong [is] inevitably a manifestation . . . of purely religious, theological, and sectarian belief".) I am willing to concede that the argument Finnis presents *is* a secular argument. Nonetheless, Finnis's secular argument is not sound. It is not an argument we should accept. Nor is it an argument that a person logically *can* accept—even a person who *wants* to accept it, who *wants* to believe that the position that homosexual sexual conduct is always wrong can be rationally vindicated—*if* that person rejects Finnis's argument that "deliberately contracepted" sexual conduct is always immoral: The Grisez–Finnis argument that "deliberately contracepted" sexual conduct is always immoral and the Grisez–Finnis argument that homosexual sexual conduct is always immoral are essentially *the same argument*.

They are the argument that any sexual conduct between two persons is always morally illicit if it cannot or does not "actualize" and "allow them to experience" their relationship as (at least in part) a *procreative* union.[122]

If Finnis's secular argument were sound, government could rely on it, for example, in refusing to extend legal recognition to same-sex unions; government would not need to rely on a religious argument—for example, the argument that God has revealed, in the Bible, that all homosexual sexual conduct is immoral. However, Finnis's argument is not sound. In the wake of Finnis's failure, one can fairly doubt that any secular argument that all homosexual sexual conduct is immoral is sound. "Once sex is no longer confined to procreative genital acts . . . , then it is no longer possible to argue that sex/love between two persons of the same sex cannot be a valid embrace of bodily selves expressing love. If sex/love is centered primarily on communion between two persons rather than on biologistic concepts of procreative complementarity, then the love of two persons of the same sex need be no less than that of two persons of the opposite sex. Nor need their experience of ecstatic bodily communion be less valuable."[123] If no secular argument that all homosexual sexual conduct is immoral is persuasive, then, as I have explained, it is difficult to see how any religious argument that all such conduct is immoral—in particular, an argument about what God has revealed, in the Bible or otherwise—should be thought sufficiently strong to ground a political choice, least of all a coercive political choice, about the morality of human conduct.

VI. A Concluding Comment (Mainly for Theologically Conservative Christians)

> *My book, O philosopher, is the nature of created things, and any time I want to read the words of God, the book is before me.*[124]

I want to emphasize, by repeating, something I said in the preface to this book. I have written this book as a Christian—in particular, as a Catholic Christian thoroughly imbued with the spirit of the Second Vatican Council; but I have written it as a Christian who is extremely wary of the God-talk in which most Christians, and many others, too often and too easily (too casually, too uncritically) engage; I have written it, that is, in the spirit of apophatic Christianity.[125] Moreover, I have written this book as one who stands between all religious nonbelievers on the one side and many religious believers—especially theologically conservative believers—on the other. Religious nonbelievers, many of whom would like to marginalize the role of religious discourse in public political debate, are the principal addressees of my argument, in chapter 2, that it is not merely permissible but important that religious arguments about the morality of human conduct be presented in public political debate. Religious nonbelievers are also the principal addressees of my argument, that in making a political choice about the morality of human conduct, legislators and others may rely on a religious argument that all human beings are sacred even if, in their view, no persuasive secular argument supports the claim about the sacredness of all human beings. By contrast, religious believers—especially Christians—are the principal addressees of my argument that in making a political choice about the morality of human conduct, especially a coercive political choice, legislators and others should not rely on a religious argument about the requirements of human well-being unless, in their view, a persuasive secular argument reaches the same conclusion about those requirements as the religious argument.

As the twentieth century draws to a close, Christians still constitute the largest religious group in the United States—although Christians in the United States are so pluralistic that I hesitate to call them a group. I want to conclude this book with a few words for those Christians most likely to view with a skeptical eye that part of my argument addressed principally to fellow Christians: theologically conservative Christians, many of whom form the base of political support for the so-called religious right in American politics today.[126] (It is precisely in that part of my argument, which concerns mainly claims about what God has revealed about the requirements of human well-being, that my wariness about God-talk is most engaged.) In speaking to such Christians, I address them not from the outside, as a religious nonbeliever speaking to religious believers. Rather, I address them as a fellow Christian. However, I hope that these concluding comments will speak, if only indirectly, to theologically conservative members of other religious traditions as well.

As I have explained, most Christians in the United States today—including Catholics, Episcopalians, and "reformed" Christians (e.g., Lutherans, Methodists, and Presbyterians)—have no basis in their religious-moral traditions for doubting that any religious argument about the requirements of human well-being is of questionable soundness unless a persuasive secular argument reaches the same conclusion about the requirements of human well-being as the religious argument. Nor, in particular, do they have a basis in their traditions for doubting that any argument about the requirements of human well-being that is grounded on a claim about what God has revealed is highly suspect if no persuasive secular route reaches the religious argument's conclusion about the requirements of human well-being. Such Christians understand that they do not have to choose between "faith" and "reason"; for them, faith and reason are not in tension, they are not incompatible. To the contrary, faith and reason are, for such Christians, mutually enriching. David Hollenbach explains: "Faith and understanding go hand in hand in both the Catholic and Calvinist views of the matter. They are not adversarial but reciprocally illuminating. As [David] Tracy puts it, Catholic social thought seeks to correlate arguments drawn from the distinctively religious symbols of Christianity with arguments based on shared public experience. This effort at correlation moves back and forth on a two-way street. It rests on a conviction that the classic symbols of Christianity can uncover meaning in personal and social existence that common sense and uncontroversial science fail to see. So it invites those outside the church to place their self-understanding at risk by what Tracy calls conversation with such 'classics.'"[127] Hollenbach then adds, following Tracy: "At the same time, the believer's self-understanding is also placed at risk because it can be challenged to development or even fundamental change by dialogue with the other—whether this be a secular agnostic, a Christian from another tradition, or a Jew, Muslim, or Buddhist."[128] I add, with an eye on an issue that has engaged me in this chapter: Whether this be a lesbian or a gay man, perhaps even a Christian lesbian or a Christian gay man, living in a lifelong, monogamous relationship of faithful love.

Predictably, some Christians—in particular, "fundamentalist" Christians and some Christian "evangelicals"—will be skeptical that an argument about the requirements of human well-being that is grounded on a claim about what God has revealed is highly suspect if there is no secular route to, if there is no argument "based on shared public experience" for, the religious argument's conclusion about the requirements of human well-being. For such Christians, faith—including faith in what God has revealed—and reason are often incompatible; in their view, human reason is too corrupted to be trusted. For example, David Smolin, a law professor who identifies himself as an evangelical Christian, has written that "even our intellectual capacities have been distorted by the effects of sin. The pervasive effects of sin suggest that creation, human nature, and human reason are often unreliable means for knowing the law of God. . . . Thus, scripture and Christian tradition have come to have a priority among the sources of knowledge of God's will. Indeed, these sources of revelation are considered a means of measuring and testing claims made on behalf of reason, nature, or creation, in order to purify these now subsidiary means of the distortive effect of sin."[129]

I want to make two points in response to theologically conservative Christians. (The points could easily be adapted to respond to theologically conservative members of other religious traditions.) First, they would do well to study Mark Noll's powerful, eloquent book, *The Scandal of the Evangelical Mind*.[130] Noll—the McManis Professor of Christian Thought at Wheaton College (Illinois), one of the foremost Christian (Protestant) colleges in the United States—is himself a committed evangelical Christian. Noll comments critically, in one chapter of his book, on the emergence of "creation science" in evangelical Christianity: "[I]f the consensus of modern scientists, who devote their lives to looking at the data of the physical world, is that humans have existed on the planet for a very long time, it is foolish for biblical interpreters to say that 'the Bible teaches' the recent creation of human beings." Noll explains: "This does not mean that at some future time, the procedures of science may shift in such a way as to alter the contemporary consensus. It means that, for people today to say they are being loyal to the Bible and to demand belief in a recent creation of humanity as a sign of obedience to Scripture is in fact being unfaithful to the Bible, which, in Psalm 19 and elsewhere, calls upon followers of God to listen to the speech that God has caused the natural world to speak. It is the same for the age of the earth and for all other questions regarding the constitution of the human race. Charles Hodges's words from the middle of the nineteenth century are still pertinent: 'Nature is as truly a revelation of God as the Bible, and we only interpret the Word of God by the Word of God when we interpret the Bible by science.'"[131] What Noll says about the proper relation between religious faith and secular inquiry into the origins of human beings is no less true about the proper relation between religious faith and secular inquiry into the well-being of human beings. "My book, O philosopher, is the nature of created things, and any time I wish to read the words of God, the book is before me."[132]

Second, theologically conservative Christians should not overlook that, as the history of Christianity discloses, sin can distort, and indeed has often distorted, "scripture and Christian tradition", not to mention what human beings believe about "scripture and Christian tradition".[133] Given their belief in the "fallenness" of human nature—which is, after all, *their* nature, too—Christians should be especially alert to this dark possibility. Smolin privileges religiously based moral arguments over secular moral arguments, but *both* sorts of arguments are, finally, human arguments. Why, then, doesn't a truly robust sense of "the distortive effect of sin" counsel that we should test religious arguments about the morality of human conduct—in particular, religious arguments about the requirements of human well-being, both those based on scripture and those based on tradition—with secular arguments about the morality of human conduct? Of course, a religious believer might well want to move in the other direction as well: She might well want to test secular arguments about the morality of human conduct with religiously based moral arguments as well as test the latter with the former. Nonetheless, Robert Audi's important point about the relative unreliability of both scripture-based and tradition-based religious arguments is worth pondering: "[I]t may be better to try to understand God through ethics than theology."[134] John Robinson, introducing a symposium on law, morality, and homosexuality, has amplified much the same point in the context of religious arguments about the (im)morality of homosexual sexual conduct:

> Jesus had little to say about human sexuality, and the canonical letters add little to the little that he is reported to have said. It is not that their authors had nothing to say about human sexuality. Paul was particularly wont to write critically of the sexual libertinism of his pagan contemporaries. No, the point is that except for a luminous passage in Paul's letter to the Ephesians, the canonical letter writers make little effort to integrate their thoughts about human sexuality into their appropriation of the Gospel message, and even in that luminous passage, modern readers can find a profoundly troubling subtext. The problem for us today is that we do not find the canonical writers making a conscious effort to distinguish what their culture told them about sex from what the Gospel told them about it. The same is true of the patristic writers and of the work of the schoolmen, all of whose work was set in an intellectual and cultural context that they themselves did not adequately distinguish from the Gospel message that they handed on to us. The result is that as we moderns come to doubt the moral propriety of patriarchalism, for example, we find that we cannot resolve that doubt by reference to scripture and tradition. They are both influenced by the same patriarchalism that we are questioning, and yet the mode of that influence is such that we would be supremely unwise to regard either Scripture or tradition as validating it for us.
>
> We find ourselves in a similar quandary as we reconsider the close nexus between morally permissible sexual activity and reproduction, a nexus that the tradition has handed down to us. Is that nexus an ineluctable implication of the Gospel message or is it an understandable but no longer relevant feature of the cultures in which that message was first articulated and later systematized? Neither scripture nor tradition answers that sort of question for us; so we must answer it for ourselves. This does not mean that we abandon scripture and tradition in our reassessment of the nexus between sex and reproduction, but it does mean that our resort to scripture and tradition has to be critical if it is to be useful.[135]

Two things bear emphasis. First, my point is not that in making a political choice about the morality of human conduct, religious believers ought not to rely on a religious argument about the requirements of human well-being. My point is that they ought not to do so *unless* a persuasive (to them) secular argument reaches the same conclusion about the requirements of human well-being as the religious argument on which they are inclined to rely. In other words, in making a political choice about the morality of human conduct—a political choice that rests on a claim about the requirements of human well-being—religious believers should rely at least partly on a secular argument about the requirements of human well-being.[136]

Second—and here I speak principally to Christians—the principle of political self-restraint I recommend here does not presuppose that in making political choices about the morality of human conduct, Christians should forget that they are Christians, that they should "bracket" their Christian identity, that they should act as if they are persons who do not have the religious beliefs that they in fact do have. (I have contended against such "bracketing" elsewhere: "One's basic moral/religious convictions are (partly) self-constitutive and are therefore a principal ground—indeed, the principal ground—of political deliberation and choice. To 'bracket' such convictions is therefore to bracket—to annihilate—essential aspects of one's very self. To participate in politics and law . . . with such convictions bracketed is not to participate as the self one is but as [someone else]."[137]) Rather, it is *because* they are Christians—it is because they are, *as Christians*, painfully aware of the fallenness, the brokenness, of human beings—that they should be extremely wary about making a political choice, least of all a coercive political choice, on the basis of a religious argument about the requirements of human well-being in the absence of any independent, corroborating secular argument.[138] (Of course, this is not to suggest that persons other than Christians can't or don't have their own powerful reasons to insist on what Christians call the fallenness or brokenness of human beings.)

It scarcely seems radical to suggest that Christians, too, like other religious believers, must be alert to the possibility that a scripture-based or a tradition-based religious argument about the morality of human conduct, no less than a secular argument, is mistaken. Christians should be at least as alert as others to this dark possibility. There is, of course, no virtue in adhering to a mistaken position, including a mistaken religious position; nor, therefore, is there any virtue in adhering to a position uncritically, so that one is unable to discern whether it is, or might be, mistaken. Indeed, uncritical adherence to a position is also, for a Christian, *unfaithful* adherence. "At any stage in history all that is available to the Church is its continual meditation on the Word of God in the light of contemporary experience and of the knowledge and insights into reality which it possesses at the time. To be faithful to that set of circumstances . . . is the charge and the challenge which Christ has given to his Church. But if there is a historical shift, through improvement in scholarship or knowledge, or through an entry of society into a significantly different age, then what that same fidelity requires of the Church is that it respond to the historical shift, such that it might be not only mistaken *but also unfaithful* in declining to do so."[139]

John Noonan's eloquent plea seems a fitting conclusion here:

> One cannot predict future changes; one can only follow present light and in that light be morally certain that some obligations will never alter. The great commandments of love of God and of neighbor, the great principles of justice and charity continue to govern all development. God is unchanging, but the demands of the New Testament are different from those of the Old, and while no other revelation supplements the New, it is evident from the case of slavery alone that it has taken time to ascertain what the demands of the New really are. All will be judged by the demands of the day in which they live. It is not within human competence to say with certainty who was or will be saved; all will be judged as they have conscientiously acted. In new conditions, with new insight, an old rule need not be preserved in order to honor a past discipline. . . .
>
> In the Church there can always be fresh appeal to Christ, there is always the possibility of probing new depths of insight. . . . Must we not, then, frankly admit that change is something that plays a role in [Christian] moral teaching? . . . Yes, if the principle of change is the person of Christ.[140]

Appendix: Judges—A Special Case?

My focus in this chapter and indeed throughout this book has been mainly on citizens, legislators, and policymaking officials in the executive branch of government, like the President of the United States or a state governor. What about the judicial branch of government? With respect to the questions addressed in this book, are judges a special case?[141] While some of us might not be accustomed to thinking of judges as policymakers, in fact some judges are sometimes policymakers, in this sense: The legal materials on which a judge must rely in deciding a case are sometimes underdeterminate; they sometimes underdetermine the resolution of the case.[142] (Because the relevant legal materials typically rule out many possible resolutions of a case even if they do not rule in just one resolution, "underdeterminate" is a more accurate term than "indeterminate".[143]) In "interpreting" such materials, the judge must decide on the direction in which the law should move.[144] Because such a decision is a decision of the judicial branch of government, it is a political—that is, a governmental—choice.

Rawls emphasizes that the ideal of public reason "applies . . . in a special way to the judiciary and above all to a supreme court in a constitutional democracy with judicial review."[145] He explains: "[T]he justices have to explain and justify their decisions as based on their understanding of the constitution and relevant statutes and precedents. Since acts of the legislative and the executive need not be justified this way, the court's special role makes it the exemplar of public reason."[146] The serious question, however, is about what a court may do when the relevant legal materials are underdeterminate, as they often are in the great constitutional cases, for example. It is Rawls's view, of course, that when the relevant materials are underdeterminate, a court must abide the strictures of the ideal of public reason. For example:

> The justices cannot, of course, invoke their own personal morality, nor the ideals and virtues of morality generally. Those they must view as irrelevant. Equally, they cannot invoke their or other people's religious or philosophical views. Nor can they cite political values without restriction. Rather, they must appeal to the political values they think belong to the most reasonable understanding of the public conception and its political values of justice and public reason. These are the values they believe in good faith, as the duty of civility requires, that all citizens as reasonable and rational might reasonably be expected to endorse.[147]

But what is a court to do when the only political values on which Rawls says the judiciary may rely are, like the relevant legal materials themselves, underdeterminate? Here, too, Rawls ignores the serious problem that the underdeterminacy of the relevant public values poses for his approach.[148]

Theories of adjudication, perhaps especially theories of constitutional adjudication, are controversial. The central question for such theories is whether in the course of shaping underdeterminate legal materials a court may sometimes rely on a controversial nonlegal norm. My position, which I have developed at length elsewhere, is that the judiciary may sometimes rely on one or more controversial nonlegal norms in shaping underdeterminate legal materials.[149] The issue now at hand is this: Is it legitimate for a court to base its shaping of underdeterminate legal materials on a controversial nonlegal norm *even if the norm is religious*? Put another way, in the context of adjudication, should nonlegal norms that are secular, some of which might be quite controversial, be privileged in relation to nonlegal norms that are religious? I can see no reason for thinking that judges are a special case. The nonestablishment norm forbids—*properly* forbids, in my view—government, including the judicial branch, to rely on a religious premise in making a choice if no plausible secular premise supports the choice.[150] If a plausible secular premise *does* support the choice, however, government, including the judicial branch, may rely on a religious premise.

Kent Greenawalt, too, concludes that sometimes—rarely, but sometimes—a court may rely on a religious premise in making a choice.[151] However, I am troubled by Greenawalt's suggestion that on the relatively rare occasion when a court may do so, the judge should not acknowledge that she has done so. According to Greenawalt, the judge should, in her opinion, conceal that she has relied on a religious premise.[152] Greenawalt does not defend this recommendation, so as I can tell, except to say that "the [court's] opinion should symbolize the aspiration of interpersonal reason and be limited to public reasons."[153] That elected officials—"politicians"—sometimes conceal that they have relied on a controversial premise in making a political choice seems, in our political culture, inevitable and, perhaps, unremarkable. But that *judges* conceal that they have relied on a controversial premise, including a religious premise, is deeply problematic, as is the proposition that they *should* do so. As an ideal matter, the parties to a case—especially the losing parties—should be informed of all the significant reasons why the case has been decided the way it has. Any other practice—including the one Greenawalt recommends—is in serious tension with the ideal of "the rule of law", which governs the process of adjudication as well as the processes of legislation (law making) and of administration (law enforcement). With respect to adjudication, the rule of law requires, on any plausible account, "that judicial decisions should be in accordance with law, issued after a fair and public hearing by an independent and impartial court, and that they should be reasoned and available to the public".[154] In what sense, and to what extent, is a judicial decision "reasoned and available to the public" if in its opinion a court conceals one of the premises on which it has consciously relied?

It does not help—and, indeed, it is to misunderstand what I have just said—to insist that a court should rely only on the relevant legal materials, perhaps supplemented where necessary by relevant public values (or, at least, by relevant values as to which there is a substantial consensus in American society). By hypothesis, we are talking about cases, however rare, in which *both* the relevant legal materials *and* the relevant public values (if there are any) are underdeterminate.

NOTES

Introduction

1. Richard N. Ostling, "In So Many Gods We Trust," Time, Jan. 30, 1995, at 72.

2. Book Note, "Religion and *Roe*: The Politics of Exclusion," 108 Harvard L. Rev. 495, 498 n. 21 (1994) (reviewing Elizabeth Mensch & Alan Freeman, The Politics of Virtue: Is Abortion Debatable? (1993)). Cf. Andrew Greeley, "The Persistence of Religion," Cross Currents, Spring 1995, at 24.

3. As a matter of political morality, secular arguments that one or another sort of human conduct is immoral, as distinct from religious arguments, are not, as such, a problematic basis of political choice. See Kent Greenawalt, "Legal Enforcement of Morality," 85 J. Criminal L. & Criminology 710 (1995).

4. Those with the principal policymaking authority and responsibility—in particular, legislators—should ask themselves whether they find a secular rationale persuasive. See ch. 1, n. 97 and accompanying text.

5. See Michael J. Perry, Love and Power: The Role of Religion and Morality in American Politics (1991).

6. See Kent Greenawalt, Religious Convictions and Political Choice (1988); Kent Greenawalt, Private Consciences and Public Reasons (1995); John Rawls, Political Liberalism (1993).

7. See Ronald Dworkin, Life's Dominion: An Argument About Abortion, Euthanasia, and Individual Freedom 25 (1993): "Some readers . . . will take particular exception to the term 'sacred' because it will suggest to them that the conviction I have in mind is necessarily a theistic one. I shall try to explain why it is not, and how it may be, and commonly is, interpreted in a secular as well as in a conventionally religious way. But 'sacred' does have ineliminable religious connotations for many people, and so I will sometimes use 'inviolable'

instead to mean the same thing, in order to emphasize the availablity of that secular interpretation."

8. See David Tracy, "Approaching the Christian Understanding of God," in Francis Schüssler Fiorenza & John P. Galvin, eds., Systematic Theology: Roman Catholic Perspectives, vol. 1, 131, 147 (1991): "[According to Christian belief,] God, the holy mystery who is the origin, sustainer, and end of all reality . . . is disclosed to us in Jesus Christ as pure, unbounded love." See also John Dominic Crossan, Jesus: A Revolutionary Biography 20 (1994): "Christian belief is (1) an act of faith (2) in the historical Jesus (3) as the manifestation of God."

9. See Michael J. Perry, "The Idea of a *Catholic* University," 78 Marquette L. Rev. 325 (1995). See also David Hollenbach, SJ, "Afterword: A Community of Freedom," in R. Bruce Douglass & David Hollenbach, SJ, eds., Catholicism and Liberalism: Contributions to American Public Philosophy 323, 337 (1994):

> For Christian believers, it is a challenge to recognize that their faith in God and the way of life it entails is a historical reality—it is rooted in historically particular scriptures and symbols and it is lived and sustained in historically particular communities. This historicity means that the task of interpreting the meaning of their faith will never be done as long as history lasts. The God in whom they place their faith can never be identified with any personal relationship, social arrangement, or cultural achievement. God transcends all of these. Though Christians believe that in Jesus Christ they have been given a definitive revelation of who this God is, they cannot claim to possess or encompass God in any of their theologies or understandings of the ultimate good of human life. Thus, in the words of Avery Dulles, "The Christian is defined as a person on the way to discovery, on the way to a revelation not yet given, or at least not yet given in final form."

(Quoting Avery Dulles, SJ, "Revelation and Discovery," in Theology and Discovery: Essays in Honor of Karl Rahner 27 (William J. Kelly, SJ, ed., 1980).) Hollenbach adds: "Because the Christian community is always on the way to the fullness of its own deepest faith, hope, and love, it must be continually open to fresh discoveries. Encounter with the other, the different, and the strange must therefore characterize the life of the church. Active participation in a community of freedom is a prerequisite to such discovery." Id. at 337.

10. On "apophatic", see 1 Oxford English Dictionary 554 (2nd ed., 1989). I concur in David Tracy's statement:

> In and through even the best speech for Ultimate Reality, greater obscurity eventually emerges to manifest a religious sense of that Reality as ultimate mystery. Silence may be the most appropriate kind of speech for evoking this necessary sense of the radical mystery—as mystics insist when they say, "Those who know do not speak; those who speak do not know." The most refined theological discourse of the classic theologians ranges widely but returns at last to a deepened sense of the

same ultimate mystery: the amazing freedom with all traditional doctrinal formulations in Meister Eckhart; the confident portrayals of God in Genesis and Exodus become the passionate outbursts of the prophets and the painful reflections of Job, Ecclesiastes, and Lamentations; the disturbing light cast by the biblical metaphors of the "wrath of God" on all temptations to sentimentalize what love means when the believer says, "God is love"; the proclamation of the hidden and revealed God in Luther and Calvin; the *deus otiosus* vision of God in the Gnostic traditions; the repressed discourse of the witches; the startling female imagery for Ultimate Reality in both the great matriarchal traditions and the great Wisdom traditions of both Greeks and Jews; the power of the sacred dialectically divorcing itself from the profane manifested in all religions; the extraordinary subtleties of rabbinic writing on God become the uncanny paradoxes of kabbalistic thought on God's existence in the very materiality of letters and texts; the subtle debates in Hindu philosophical reflections on monism and polytheism; the many faces of the Divine in the stories of Shiva and Krishna; the puzzling sense that, despite all appearances to the contrary, there is "nothing here that is not Zeus" in Aeschylus and Sophocles; the terror caused by Dionysius in Euripides' *Bacchae*; the refusal to cling even to concepts of "God" in order to become free to experience Ultimate Reality as Emptiness in much Buddhist thought; the moving declaration of that wondrous clarifier Thomas Aquinas, "All that I have written is straw; I shall write no more"; Karl Rahner's insistence on the radical incomprehensibility of both God and ourselves understood through and in our most comprehensible philosophical and theological speech; . . . the "God beyond God" language of Paul Tillich and all theologians who acknowledge how deadening traditional God-language can easily become; the refusal to speak God's name in classical Judaism; the insistence on speaking that name in classical Islam; the hesitant musings on the present-absent God in Buber become the courageous attempts to forge new languages for a new covenant with God in the post-*tremendum* theologies of Cohen, Fackenheim, and Greenberg. There is no classic discourse on Ultimate Reality that can be understood as mastering its own speech. If any human discourse gives true testimony to Ultimate Reality, it must necessarily prove uncontrollable and unmasterable.

David Tracy, Plurality and Ambiguity: Hermeneutics, Religion, Hope 108-09 (1987). See also Martin Buber, quoted in Hans Küng, Does God Exist? An Answer for Today 508 (1980):

["God"] is the most loaded of all words used by men. None has been so soiled, so mauled. But that is the very reason I cannot give it up. Generations of men have blamed this word for the burdens of their troubled lives and crushed it to the ground; it lies in the dust, bearing all their burdens. Generations of men with their religious divisions

> have torn the word apart; they have killed for it and died for it; it bears all their fingerprints and is stained with all their blood. Where would I find a word to equal it, to describe supreme reality? If I were to take the purest, most sparkling term from the innermost treasury of the philosophers, I could capture in it no more than a noncommittal idea, not the presence of what I mean, of what generations of men in the vastness of their living and dying have venerated and degraded. . . . We must respect those who taboo it, since they revolt against the wrong and mischief that were so readily claimed to be authorized in the name of God; but we cannot relinquish it. It is easy to understand why there are some who propose a period of silence about the "last things," so that the misused words may be redeemed. But this is not the way to redeem them. We cannot clean up the term "God" and we cannot make it whole; but, stained and mauled as it is, we can raise it from the ground and set it above an hour of great sorrow.

For a powerful and acclaimed recent reflection on God-talk, see Elizabeth A. Johnson, She Who Is: The Mystery of God in Feminist Theological Discourse (1992).

11. The subject of this book—religion in politics—is very large. A single book can do no more than approach the subject from some of the relevant perspectives. As I have indicated in this preface, I approach the subject from two distinct but related perspectives: the perspective of American constitutional law and the perspective of political morality. Other important, complementary perspectives—those of history and of the social sciences, for example—are beyond my expertise.

I do not address, in this book, the various practical questions one might ask about the proper role of religious institutions and personnel in American political life—for example, should "clergy, churches, and other religious organizations engage in ordinary political activities, such as educational campaigns, lobbying, demonstrations, and attempts to put strong electoral pressure on officials?" Greenawalt, Private Consciences and Public Reasons, n. 6, at 165. (Greenawalt addresses such questions in ch. 15 of his book.)

Chapter One

1. For the reader interested in something I do not provide in this book—commentary on the details of the Supreme Court's principal decisions and doctrines regarding religious liberty—there are numerous recent works, including: Jesse H. Choper, Securing Religious Liberty: Principles for Judicial Interpretation of the Religion Clauses (1995); Frederick Mark Gedicks, The Rhetoric of Church and State: A Critical Analysis of Religion Clause Jurisprudence (1995); Michael W. McConnell, "Religious Freedom at a Crossroads," 59 U. Chicago L. Rev. 115 (1992).

2. Only one other provision of the Constitution mentions religion: Article VI states that "no religious Test shall ever be required as a Qualification to any Office or public Trust under the United States."

3. See Michael W. McConnell, "Accommodation of Religion: An Update and Response to the Critics," 60 George Washington L. Rev. 685, 690 (1992): "The government may not 'establish' religion and it may not 'prohibit' religion." McConnell explains, in a foonote attached to the word "establish", that "[t]he text [of the First Amendment] states the 'Congress' may make no law 'respecting an establishment' of religion, which meant that Congress could neither establish a national church nor interfere with the establishment of state churches as they then existed in the various states. After the last disestablishment in 1833 and the incorporation of the First Amendment against the states through the Fourteenth Amendment, this 'federalism' aspect of the Amendment has lost its significance, and the Clause can be read as forbidding the government to establish religion." Id. at 690 n. 19.

4. John Hart Ely, Democracy and Distrust: A Theory of Judicial Review 105 (1980).

5. For a thoughtful argument that came to my attention after I had drafted this chapter—an argument for extending the command of the First Amendment beyond Congress and even beyond lawmaking generally to all public officials—see John H. Garvey, What Are Freedoms For?, ch. 14 (1996).

6. See Michael J. Perry, The Constitution in the Courts: Law or Politics? 63–68 (1994). See also Randy E. Barnett, ed., 1 The Rights Retained by the People: The History and Meaning of the Nineteenth Amendment (1989); Randy E. Barnett, ed., 2 The Rights Retained by the People: The History and Meaning of the Nineteenth Amendment (1993); Thomas B. McAffee, "The Original Meaning of the Ninth Amendment," 90 Columbia L. Rev. 1215 (1990); Thomas B. McAffee, "The Bill of Rights, Social Contract Theory, and the Rights 'Retained' by the People," 16 Southern Illinois L. J. 267 (1992); Thomas B. McAffee, "Prolegomena to a Meaningful Debate on the 'Unwritten Constitution' Thesis," 61 U. Cincinnati L. Rev. 107 (1992).

7. See Akhil Reed Amar, "The Bill of Rights and the Fourteenth Amendment," 101 Yale L. J. 1193 (1992); Michael Kent Curtis, No State Shall Abridge: The Fourteenth Amendment and the Bill of Rights (1986). See also Richard L. Aynes, "On Misreading John Bingham and the Fourteenth Amendment," 103 Yale L. J. 57 (1993). On the Fourteenth Amendment's protection of the free exercise of religion, see Kurt T. Lash, "The Second Adoption of the Free Exercise Clause: Religious Exemptions Under the Fourteenth Amendment," 88 Northwestern U. L. Rev. 1106 (1994).

8. See, e.g., Raoul Berger, The Fourteenth Amendment and the Bill of Rights (1989).

9. See Jay S. Bybee, "Taking Liberties with the First Amendment: Congress, Section 5, and the Religious Freedom Restoration Act," 48 Vanderbilt L. Rev. 1539 (1995). Some have argued that although the Fourteenth Amendment was or might have been meant to "incorporate" the First Amendment's ban on gov-

ernment prohibiting the free exercise of religion, thereby making it applicable to the states, it is implausible to believe that the Amendment was meant to incorporate the ban on government establishing religion. See, e.g., Daniel O. Conkle, "Toward a General Theory of the Establishment Clause," 82 Northwestern U. L. Rev. 1115, 1136–42 (1988); Note, "Rethinking the Incorporation of the Establishment Clause: A Federalist View," 105 Harvard L. Rev. 1700 (1992). According to Steven Smith, however, with respect to the question of incorporation, there is no basis for distinguishing between the First Amendment's ban on prohibiting the free exercise of religion and its ban on establishing religion. See Steven D. Smith, Foreordained Failure: The Quest for a Constitutional Principle of Religious Freedom, chs. 2–3 (1995). (But cf. Lash, n. 7.) According to Bybee, there is no basis for distinguishing between the First Amendment's bans concerning religion and its bans concerning speech and press. See Bybee, supra this note, at 1555–66.

10. See Cantwell v. Connecticut, 310 U.S. 296 (1940) (Fourteenth Amendment makes ban on government prohibiting the free exercise of religion applicable to states); Everson v. Board of Education, 330 U.S. 1 (1947) (Fourteenth Amendment makes ban on government establishing religion applicable to states). For reasons not important here, the Court's incorporation position, unlike that of the scholars cited in the preceding footnote, relies not on the privileges or immunities clause of section one, but on the due process clause. See, e.g., McIntyre v. Ohio Elections Commission, 115 S.Ct. 1511, 1514 n. 1 (1995): "The term 'liberty' in the Fourteenth Amendment to the Constitution makes the First Amendment applicable to the States."

11. See Michael J. Perry, "What Is 'The Constitution'?," in Lawrence A. Alexander, ed., Constitutionalism: Philosophical Foundations (forthcoming, Cambridge University Press, 1997).

12. The Preamble to the Constitution of the United States states: "We the people of the United States, in Order to form a more perfect Union, establish Justice, insure domestic Tranquility, provide for the common defence, promote the general welfare, and secure the Blessings of Liberty to ourselves and our Posterity, do ordain and establish this Constitution for the United States of America."

13. See Perry, "What is 'The Constitution'?," n. 11.

14. This is not to say that there is no serious controversy about how to justify maintaining the nonestablishment and free exercise norms as constitutional norms. Cf. id.

15. For a list of typical religious practices, see, in the appendix to this chapter, Article 6 of the Declaration on the Elimination of All Forms of Intolerance and of Discrimination Based on Religion or Belief.

What is a "religious" belief? See this ch., pp. 31–32 Cf. George C. Freeman, III, "The Misguided Search for the Constitutional Definition of 'Religion'," 71 Georgetown L. J. 1519 (1983).

16. In a recent paper, written under the supervision of my colleague, Steven Calabresi, Peter Braffman has discussed the likely original understanding of the word "free" in the free exercise language of the First Amendment. See Peter A.

Braffman. "The Original Understanding of the Free Exercise Clause," unpublished ms. at 37–38 (May 1995):

> The first definition provided by [Samuel] Johnson's dictionary, which was uniformly given by other contemporary dictionaries, is "liberty." The word "liberty" was understood then as it is now, as "freedom," a freedom as "opposed to slavery," a freedom as "opposed to necessity." . . . To understand "free" in this sense, is to more readily understand the place of the right of "exercise of religion" within the First Amendment. Just like the "freedom of speech" and the "freedom of the press," the "freedom" of exercise of religion can be understood as a fundamental right which precedes the social compact. . . . It is not a right "granted" or "permitted" by the sovereign, but rather an unalienable right guaranteed and protected by constitutional government. Furthermore, it would not be a right to licentiousness which is not bound by law, but rather a right to be free from unjust restraints.

17. See, e.g., Lundman v. McKown, 530 N.W.2d 807 (Minnesota 1995). See also Caroline Frasier, "Suffering Children and the Christian Science Church," Atlantic Monthly, April 1995, at 105.

18. Discussing "a distinction that is implicit in the idea of 'persecution'" and that the Court in *Church of the Lukumi Babalu Aye, Inc. v. City of Hialeah* (113 S.Ct. 2217 (1993)) "repeatedly tried to articulate", Steve Smith has explained: "The distinction is between measures that 'target' a religion on *religious grounds* and because it is objectionable *as* a religion and, on the other hand, measures that 'target' a religion only by prohibiting a practice of the religion that is objectionable *on independent or nonreligious grounds*." Steven D. Smith, "Free Exercise Doctrine and the Discourse of Disrespect," 65 U. Colorado L. Rev. 519, 563 (1994). See id. at 563–68.

19. Employment Division, Department of Human Resources of Oregon v. Smith, 494 U.S. 872, 877–78 (1990).

20. To favor one or more religions is not necessarily to take prohibitory action disfavoring one or more religious practices. Action that violates the nonestablishment norm, therefore, does not necessarily violate the free exercise norm.

21. Moreover, Article 6 states, in relevant part: "All powers of government, legislative, executive, and judicial, derive, *under God*, from the people, whose right it is to designate the rulers of the State and, in the final appeal, to decide all questions of national policy, according to the requirements of the common good." (Emphasis added.) And Article 44 of the Constitution states, in relevant part: "The State acknowledges that the homage of public worship is due to Almighty God. It shall hold His Name in reverence, and shall respect and honor religion." (Article 40 states that "[t]he publication or utterance of *blasphemous*, seditious, or indecent matter is an offense which shall be punishable in accordance with law." (Emphasis added.)) On "religion in the Preamble}, see Gerard Hogan & G. F. Whyte, J.M. Kelly's The Irish Constitution 6–7 (3rd ed., 1994).

Although it affirms Christianity, the Irish Constitution explicitly disallows the "endowing" of any religion. Article 44.2.1 states: "The State guarantees not to endow any religion."

22. The idea of human rights is that each and every human being is sacred, and that therefore there are certain things that ought not to be done to any human being and certain other things that ought to be done for every human being. See Michael J. Perry, The Idea of Human Rights: Four Inquiries (forthcoming, 1998).

23. In correspondence, Gerry Whyte of the Trinity College (Dublin) School of Law (and co-author of J.M. Kelly's The Irish Constitution, n. 21) has called my attention to "the sectarian nature, in the context of the Christian tradition, of the Preamble. Which Christian denomination was sustained by Jesus Christ 'through centuries of trial' in Irish history? And what does that imply about Christ's attitude to that denomination which was responsible for such oppression? Criticism of the Preamble is not so much that it endorses Christianity above other world religions but rather that it is sectarian within the Christian tradition." Letter to Michael J. Perry, July 12, 1994. Nonetheless, the existence in the Irish Constitution of the clause to which Professor Whyte refers—"who sustained our fathers through centuries of trial"—is understandable, given the extreme brutality, including religious oppression, endured by the majority Catholic population during the several centuries in which they were a colonized people. In any event, it is difficult to see how the existence of the clause in the Irish Constitution violates anyone's human rights. (Incidentally, anyone who thinks that the Catholic Church calls the shots in Ireland knows little about contemporary Ireland. See Colm Tóibín, "Dublin's Epiphany: Letter from Ireland," New Yorker, Apr. 3, 1995, at 45.)

24. Article 44 also states that "[l]egislation providing State aid for schools shall not discriminate between schools under the management of different religious denominations, *nor be such as to effect prejudicially the right of any child to attend a school receiving public money without attending religious instruction at that school*." (Emphasis added.)

25. There *is* a problem with the Irish situation, however. The presidential oath required by Article 12 of the Irish Constitution reads as follows: "In the presence of Almighty God I do solemnly and sincerely promise and declare that I will maintain the Constitution of Ireland and uphold its laws, that I will fulfill my duties faithfully and conscientiously in accordance with the Constitution and law, and that I will dedicate my abilities to the service and welfare of the people of Ireland. May God direct and sustain me." The authors of J.M. Kelly's The Irish Constitution inform us that "the Human Rights Committee of the United Nations recently indicated that the religious flavour of the Presidential oath violated Article 18 of the International Covenant on Civil and Political rights on freedom of thought, conscience and religion." Hogan & Whyte, n. 21, at 85 (citing The Irish Times, July 15, 1993). Presumably the same problem attends the judicial oath required by Article 34, which has the very same "religious flavour". The judicial oath begins with "In the presence of Almighty God" and ends with "May God direct and sustain me."

26. Brian Barry, Justice as Impartiality 165 n. *c* (1995).

27. Again, because to favor one or more religions (as such) is not necessarily to take prohibitory action disfavoring one or more religious practices (as such), action that violates the nonestablishment norm does not necessarily violate the free exercise norm.

28. Commenting on the nonestablishment norm, Doug Laycock has said something in which I concur: "[I]f I had to give up one of the rights in the First Amendment, this is the one I would give up. A rule against government persuasion or influence is less critical than a rule against government coercion. In terms of history, in terms of comparative law, in terms of what the rest of the world does, the Establishment Clause is an extraordinary protection. We would probably still be a free society without it. But I at least would mourn the loss. Repeal of our protection against religious persuasion by government would be a serious loss. . . ." Douglas Laycock, "The Benefits of the Establishment Clause," 42 DePaul L. Rev. 373, 379 (1992). For a helpful comparative perspective, suggesting what some of the losses might be were we Americans to abandon the nonestablishment norm, see Richard S. Kay, "The Canadian Constitution and the Dangers of Establishment," 42 DePaul L. Rev. 361 (1992).

29. Derek H. Davis, "Assessing the Proposed Religious Equality Amendment," 37 J. Church & State 493, 507–08 (1995). See also Marsh v. Chambers, 463 U.S. 783, 804 & n. 16 (1983) (Brennan, J., joined by Marshall, J., dissenting). As Doug Laycock has explained:

> It is not good for religion to have government engaged in religious rituals. Government by its sheer size and prominence will have a disproportionate influence on the kinds of rituals that are exercised and on public perception of what are appropriate rituals. The result will not be pretty. Government-sponsored religion is theologically and liturgically thin. It is politically compliant. It is supportive of incumbent administrations. In intolerant communities it inevitably tends to impose the majority's forms, rituals, and terminology on everybody. In tolerant communities, efforts to be all-inclusive inevitably lead to desacralization, to the least common denominator, to a secular incarnation with plastic reindeer, to Christmas and Chanukah mushed together as the Winter Holidays. By stripping all the specific elements of different faiths and denominations and attempting to keep all the common elements that all faiths share, tolerant governments produce a mishmash that no faith can accept or believe in. It has always been a great puzzle to me why certain elements of the religious community invest so much effort in demanding that government model bad religion in this way. There are serious costs to the government religious observances.

Laycock, "The Benefits of the Establishment Clause," n. 28, at 380–81.

30. Id. at 380. The point that there is no need for government to discriminate in favor of religion applies to more than just religious rituals.

31. Davis, n. 29, at 508.

32. See, e.g., Church of the Lukumi Babalu Aye, Inc. v. City of Hialeah, 113 S.Ct. 2217 (1993) (unanimous decision). There is no reason to doubt that the Court's two newest members, Ruth Ginsburg (who replaced Byron White) and Stephen Breyer (who replaced Harry Blackmun), accept that the free exercise norm is, whatever else it is, an antidiscrimination provision.

33. For a useful discussion of the Court's pragmatic but compromising refusal to invalidate the national motto ("In God We Trust"), the Pledge of Allegiance ("one nation, Under God"), and the like—a discussion that makes clear that the Court's refusal does not flow from, but compromises, its general nonestablishment philosophy—see Carl H. Esbeck, "A Restatement of the Supreme Court's Law of Religious Freedom: Coherence, Conflict, or Chaos?," 70 Notre Dame L. Rev. 381, 603–04 n. 82 (1995).

34. For a development of the point, see Lee v. Weisman, 505 U.S. 577, 592 et seq. (1992).

35. See Steven G. Gey, "Religious Coercion and the Establishment Clause," 1994 U. Illinois L. Rev. 463 (1994). See also McConnell, "Religious Freedom at a Crossroads," n. 1, at 157–65. According to McConnell, "an emphasis on coercion *could* tend toward acquiescence in more subtle forms of governmental power. . . . If interpreted strictly, the coercion test would increase the power and discretion of majoritarian institutions over matters of religion." Id. at 159.

36. If one's reasons for wanting not to be present at a religious service are religious reasons, then there might be a free exercise problem as well as a nonestablishment one. But whether or not there is a free exercise problem, there is a nonestablishment problem—and it is fatal.

37. For the Court's most recent discussion, see Lee v. Weisman, 505 U.S. 577 (1992). See also Wallace v. Jaffree, 472 U.S. 38 (1985). But cf. Marsh v. Chambers, 463 U.S. 783 (1983) (Nebraska legislature's practice of paying chaplain to begin its sessions with a prayer not violative of the nonestablishment norm). *Marsh* was wrongly decided, for the reasons given by Justice Brennan in a dissenting opinion joined by Justice Marshall (id. at 795–822).

38. As distinct from government permitting a private party to display a religious symbol on public property. See n. 40.

39. See, e.g., Stone v. Graham, 449 U.S. 39 (1980) (per curiam) (posting of Ten Commandments in public school held to violate nonestablishment norm).

40. See Lynch v. Donnelly, 465 U.S. 668 (1984); Allegheny County v. Greater Pittsburgh ACLU, 492 U.S. 573 (1989).

My focus here has been on goverment display of a religious symbol. A related issue concerns government permitting a private party to display a religious symbol on public property. The Supreme Court recently listed three "factors that we consider[] determinative . . .[:] The State did not sponsor respondents' expression, the expression was made on government property that had been opened for speech, and permission was requested through the same application process and on the same terms required of other private groups." Capitol Square Review and Advisory Board v. Pinette, 115 S.Ct. 2440, 2447 (1995). The Court's approach seems right: If (and only if) the Court's three criteria are satisfied, the inference is sound that goverment's grant of permission was not based on the view that one or more religions (or religious practices or tenets) are, as such, better than one or more other religions or than no religion at all. Cf. Lamb's Chapel v. Center Moriches Union Free School District, 508 U.S. 384 (1993); Widmar v. Vincent, 454 U.S, 263 (1981).

41. Epperson v. Arkansas, 393 U.S. 97, 98–99 (1968).

42. Id. at 98.

43. Id. at 107–08.

44. Edwards v. Aguilard, 482 U.S. 578, 581 (1987).

45. Id. at 631 (Scalia, J., dissenting, joined by Rehnquist, C.J.). Later in his dissent, Justice Scalia wrote: "Perhaps what the Louisiana Legislature has done is unconstitutional because there *is* no [scientific] evidence [in support of 'creation science'], and the scheme they have established will amount to no more than a presentation of the Book of Genesis. But we cannot say that on the evidence before us in the summary judgment context, which includes ample uncontradicted testimony that 'creation science' is a body of scientific knowledge rather than revealed belief. *Infinitely less* can we say (or should we say) that the scientific evidence for evolution is so conclusive that no one could be gullible enough to believe that there is any real scientific evidence to the contrary, so that the legislature's stated purpose must be a lie. Yet that illiberal judgment, that *Scopes*-in-reverse, is ultimately the basis on which the Court's facile rejection of the Louisiana Legislature's purpose must rest." Id. at 634.

46. Supreme Court majorities have been fairly relaxed about government aid to religiously affiliated colleges and universities. See, e.g., Roemer v. Board of Public Works, 426 U.S. 736 (1976). By contrast, Court majorities have been have been quite agitated about government aid to religiously affiliated elementary and secondary schools. The Court's rulings with respect to the latter sort of government aid (which rulings are, in a word, chaotic; see n. 46), are refracted in a recent opinion by the United States Court of Appeals for the Ninth Circuit: Walker v. San Francisco Unified School District, 46 F.3d 1449 (9th Cir. 1995).

47. See Choper, n. 1, at 174–76. This is not to deny that some Supreme Court justices have staked out a reasonably consistent position in the cases in question. Nonetheless, the position of the Court as a whole rests on no nonarbitrary distinction between the contexts in which the Court has invalidated government aid to religiously affiliated elementary or secondary schools and the contexts in which it has not.

48. See Rosenberger v. Virginia, 115 S.Ct. 2510 (1995) (majority opinion, delivered by Kennedy, J., joined by Renhquist, C.J., & O'Connor, Scalia, & Thomas, JJ.). (See also Board of Education of Kiryas Joel v. Grumet, 114 S.Ct. 2481, 2488 (O'Connor, J., concurring in part and concurring in the judgment); 2505 (Kennedy, J., concurring in the judgment); 2525 (Scalia, J., joined by Rehnquist, C.J., & Thomas, J., dissenting).) In *Rosenberger*, the Court stated: "The governmental program here is neutral toward religion. There is no suggestion

that the University created it to advance religion or adopted some ingenious device with the purpose of aiding a religious cause." 115 S.Ct. at 2522. As Justice Thomas said in a concurring opinion in *Rosenberger*, "The [Establishment] Clause does not compel the exclusion of religious groups from government benefit programs that are generally available to a broad class of participants." Id. at 2532 (Thomas, J., concurring). Commenting on an historical argument, Justice Thomas wrote:

> [T]he history cited by the dissent cannot support the conclusion that the Establishment Clause "categorically condemn[s] state programs directly aiding religious activity" when that aid is part of a neutral program available to a wide array of beneficiaries. Even if Madison believed that the principle of nonestablishment of religion precluded governmental financial support for religion *per se* (in the sense of government benefits specifically targeting religion), there is no indication that at the time of the framing he took the dissent's extreme view that the government must discriminate against religious adherents by excluding them from more generally available financial subsidies. . . . The dissent identifies no evidence that the Framers intended to disable religious entities from participating on neutral terms in even-handed government programs. The evidence that does exist points in the opposite direction. . . .

Id. at 2530, 2533.

49. Thus, I disagree with Kathleen Sullivan, who believes that an objection to one's tax monies being spent to aid religious activities merits constitutional status. See Kathleen M. Sullivan, "Religion and Liberal Democracy," 59 U. Chicago L. Rev. 195, 208–14 (1992).

50. See the preface to this book, nn. 1–2 and accompanying text.

51. For President Clinton's development of the point, see "Text of President Clinton's Memorandum on Religion in Schools," New York Times, July 13, 1995, at A10.

52. Everson v. Board of Education, 330 U.S. 1, 18 (1947). It is far from clear that, even accepting the principle stated by the Court, *Everson* was rightly decided. See id. at 18 (Jackson, J., dissenting); id. at 28 (Rutledge, J., joined by Frankfurter, Jackson, & Burton, JJ., dissenting).

53. For an extended presentation and defense of the accommodation position, see Michael W. McConnell, "The Origins and Historical Understanding of Free Exercise of Religion," 103 Harvard L. Rev. 1309 (1993); McConnell, "Accommodation of Religion," n. 3. But see William P. Marshall, "The Case Against the Constitutionally Compelled Free Exercise Exemption," 40 Case Western Reserve L. Rev. 357, 375–79 (1989–90); Gerard V. Bradley, "Beguiled: Free Exercise Exemptions and the Siren Song of Liberalism," 20 Hofstra L. Rev. 245 (1991); Philip A. Hamburger, "A Constitutional Right of Religious Exemption: An Historical Perspective," 60 George Washington L. Rev. 915 (1992).

54. Controversial questions lurk here—questions in the domain that we in

the legal academy conventionally refer to as "constitutional theory". Although I cannot address those questions here, I have recently addressed many of them elsewhere. See Perry, "What Is 'The Constitution'?," n. 11.

55. See McConnell, "The Origins and Historical Understanding of the Free Exercise Clause," n. 53. McConnell's position has been criticized. See, e.g., Bradley, n. 53; Hamburger, n. 53. However, Peter Braffman has presented a powerful argument that McConnell's position is closer to the truth than are the views of those who have criticized his position. See Braffman, n. 16, at 7–18. According to Braffman, "protecting 'free exercise of religion' meant protecting those religiously motivated actions which did not disrupt the public peace." Id. at 81. Braffman elaborates: "The public peace did not refer to all laws of the land, but rather to a limited category of laws which prevented overt acts of violence that disrupted civil society. As such, religiously motivated conduct could conflict with—and be exempted from—general laws so long as that conduct did not violate the peace." Id. at 96.

56. See Lash, n. 7.

57. Cf. Bybee, n. 9.

58. As I said, controversial questions of constitutional theory lurk here. See Perry, "What Is 'The Constitution'?", n. 11.

59. Compare McConnell, "The Origins and Historical Understanding of Free Exercise of Religion," n. 53, with Bradley, n. 53, and Hamburger, n. 53. With respect to the Fourteeth Amendment and the question of the "incorporation" of the First Amendment, compare Lash, n. 7, with Bybee, n. 9.

60. Employment Division, Department of Human Resources of Oregon v. Smith, 494 U.S. 872, 878 (1990). After the Court's decision in *Smith*, which involved Oregon's failure to exempt the religous use of peyote from its ban on the ingestion of hallucinogenic substances, Oregon passed a law making it lawful to use peyote in connection "with the good faith practice of a religious belief" or association "with a religious practice." Oregon Revised Statutes § 475.992(5)(a)&(b) (1991).

61. Three members of the Court (Justices Brennan, Marshall, and Blackmun) joined a statement by Justice O'Connor that the accommodation position should not have been rejected by the majority. See 494 U.S. at 892–903.

62. Public Law 103–141, 42 U.S. Code 2000bb.

63. On RFRA, see Thomas C. Berg, "What Hath Congress Wrought? An Interpretive Guide to the Religious Freedom Restoration Act," 39 Villanova L. Rev. 1 (1994); Scott C. Idleman, "The Religious Freedom Restoration Act: Pushing the Limits of Legislative Power," 73 Texas L. Rev. 247 (1994); Douglas Laycock & Oliver S. Thomas, "Interpreting the Religious Freedom Restoration Act," 73 Texas L. Rev. 209 (1994); "Symposium: The Religious Freedom Restoration Act," 56 Montana L. Rev. 1–294 (1995). For an important judicial statement about what it means for government to "substantially burden a person's exercise of religion", see Mack v. O'Leary, 80 F.3rd 1175 (7th Cir. 1996).

Did Congress exceed the scope of its constitutional power in enacting the Religious Freedom Restoration Act? So far, three federal courts have answered in the negative. See Flores v. City of Boerne, Texas, 73 F.3d 1352 (5th Cir. 1996); Sasnett v. Sullivan, 65 USLW 2115 (7th Cir. 1996); Belgard v. Hawaii, 883 F.Supp. 510 (D Hawaii 1995). Scholarly opinion is divided. See Bybee, n. 9 (unconstitutional); Daniel O. Conkle, "The Religious Freedom Restoration Act: The Constitutional Significance of an Unconstitutional Statute," 56 Montana L. Rev. 39 (1995) (unconstitutional); Christopher L. Eisgruber & Lawrence G. Sager, "Why the Religious Freedom Restoration Act Is Unconstitutional," 69 New York University L. Rev. 437 (1995) (unconstitutional); Eugene Gressman & Angela C. Carmella, "The RFRA Revision of the Free Exercise Clause," 57 Ohio State L. J. 65 (1996) (unconstitutional); Marci A. Hamilton, "The Religious Freedom Restoration Act: Letting the Fox into the Henhouse Under Cover of Section Five of the Fourteenth Amendment," 16 Cardozo L. Rev. 357 (1994) (unconstitutional); Douglas Laycock, "RFRA, Congress, and the Ratchet," 56 Montana L. Rev. 145, 152–70 (1995) (constitutional); Bonnie I. Robin-Vergeer, "Disposing of the Red Herrings: A Defense of the Religious Freedom Restoration Act," 69 Southern California L. Rev. 589 (1996) (constitutional). Robin-Vergeer's excellent article is, in my view, persuasive.

64. See Ian Brownlie, ed., Basic Documents on Human Rights 125 (1992).

65. Id. at 132.

66. See id. at 25, 27.

67. See id. at 110.

68. See id. at 501.

69. See id. at 330.

70. See, e.g., Suzanna Sherry, "Lee v. Weisman: Paradox Redux," 1993 Supreme Court Review 123, 136–50; William P. Marshall, "The Religious Freedom Restoration Act: Establishment, Equal Protection, and Free Speech Concerns," 56 Montana L. Rev. 227, 237–42 (1995); Christopher L. Eisgruber & Lawrence G. Sager, "Why the Religious Freedom Restoration Act Is Unconstitutional," n. 63.

One way to try to avoid this sobering argument is by requiring an accommodation of a religious practice if, *but only if*, the refusal to accommodate the practice—by exempting it from an otherwise applicable regulatory restraint—seems based on the view that as religious practice, the practice is inferior to one or more other religious practices. A refusal to accommodate based on such a view would violate the free exercise norm, which forbids government to take any action, impeding a religious practice, based on the view that the practice is, as such, inferior to another religious or nonreligious practice or to no practice at all. Imagine, for example, that a state bans the consumption of alcoholic beverages but, like the United States during Prohibition, exempts the consumption of wine in the Christian sacrament of the Eucharist. Imagine further that the state also bans the ingestion of hallucinogenic substances and refuses to exempt the ingestion of peyote in a Native American religious ceremony. This state of affairs supports a rebuttable inference that the state is devaluing the religious use of peyote, as religious practice, relative to the religious use of wine, which the state is accommodating. If a prima facie showing has been made that the state is devaluing the religious use of peyote, shouldn't the state be required to exempt the religious use of peyote—unless it can demonstrate that such an exemption, unlike the exemption for the religious use of wine, would seriously compromise an important public interest? After all, unless the state can make that showing, the rebuttable inference stands that the failure to exempt is a discriminatory act in violation of the free exercise norm—the free exercise norm, that is, *understood only as an antidiscrimination provision*. (See McConnell, "Accommodation of Religion," n. 3, at 707: "[I]f the Prohibition had exempted wine used for a eucharistic mass but not wine used for a Jewish Seder, that would have been discriminatory. If Congress exempted Quakers and Mennonites from the draft, but not members of other churches who entertain similar convictions about participation in war, that would be discriminatory. Discrimination of this sort would require powerful justification.") Moreover, it should not matter to the result in the peyote case that the state does not ban the consumption of alcoholic beverages, so long as there is little doubt that if the state *did* ban their consumption, the state *would* exempt the religious use of wine.

Alas, there is a problem with the argument that if government does or would favor, by accommodating, a religious practice—e.g., the religious use of wine—it must (absent justification) accommodateother, similar religious practices—e.g., the religious use of peyote: Why should those who reject the accommodation position accept the implicit premise that it is permissible for government to accommodate a religious practice as such—like the religious use of wine—if government wants to do so? The rejectionists' argument against the accommodation position, after all, is that for government to favor, by accommodating, religious practice as such is for government, in violation of the nonestablishment norm, to take action based on the view that religious practice is, as such, better or more valuable than nonreligious practice.

71. Professors Eisgruber and Sager have raised "the possibility that a constitutional privilege for religion could be rehabilitated if it were generalized to include a wider range of human commitments and thus avoid the complaints of sectarianism or partisanship. The idea would be to privilege all acts of conscience, not merely those that are rooted in a conventionally religious system of belief." Christopher L. Eisgruber & Lawrence G. Sager, "The Vulnerability of Conscience: The Constitutional Basis for Protecting Religious Conduct," 61 U. Chicago L. Rev. 1245, 1268 (1994). Laura Underkuffler has defended and pursued the doctrinal implications of the position that individual conscience should be protected in the name of religious freedom without regard to whether the conscience is "religious" in any conventional sense. See Laura Underkuffler, "The Separation of the Religious and the Secular: A Foundational Challenge to First Amendment Theory," 36 William & Mary L. Rev. 837 (1995).

72. Scott Idleman has anticipated such an suggestion. See Scott C. Idleman, "The Sacred, the Profane, and the Instrumental: Valuing Religion in the Culture of Disbelief," 142 U. Pennsylvania L. Rev. 1313, 1375–76 (1994).

73. For a thoughtful, provocative discussion of why constitutional doctrine regarding religious freedom should be reoriented to "focus . . . on the protection

of individual conscience" without regard to whether the conscience is "religious" in any conventional sense—and of what the reoriented constitutional doctrine would look like—see Underkuffler, n. 71. (The quoted language appears in id. at 961.)

74. An exemption for the refusal of Christian Science parents to seek readily available lifesaving medical care for their gravely ill child would obviously compromise an important public interest. See, e.g., Lundman v. McKown, 530 N.W.2d 807 (Minnesota 1995). But see Stephen L. Carter, "The Power of Prayer, Denied," New York Times, Jan. 31, 1996, at A11.

75. Church of the Lukumi Babalu Aye, Inc. v. City of Hialeah, 113 S.Ct. 2217, 2218 (1993).

76. 494 U.S. 872 (1990).

77. Smith, "Free Exercise Doctrine and the Discourse of Disrespect," n. 18, at 575–76. Smith discusses, and applauds, the Court's performance in *Wisconsin v. Yoder* (406 U.S. 205 (1972)) at various points in his excellent article.

78. Accepting the accommodation position back into the constitutional law of the United States might not mean very much as a practical matter, however, given the Supreme Court's depressing record of failing to take free exercise claims very seriously—even when the accommodation position was a part of the constitutional law of the United States. See John T. Noonan, Jr., "The End of Free Exercise," 42 DePaul L. Rev. 567 (1992). But cf. Laycock, "The Benefits of the Establishment Clause," n. 28, at 376: "Some of the time, judicial review will do some good. Judges did nothing for the Mormons, but they may have saved the Jehovah's Witnesses and the Amish. If judges can save one religious minority a century, I consider that ample justification for judicial review in religious liberty cases."

When would an exemption seriously compromise an important public "interest? The Texas statute ruled unconstitutional by the Supreme Court in *Texas Monthly, Inc. v. Bullock* provides an exemplary illustration. The statute exempted from sales and use taxes "[p]eriodicals that are published or distributed by a religious faith and that consist wholly of writings promulgating the teachings of the faith and books that consist wholly of writings sacred to a religious faith." 489 U.S. 1, 5 (1989). A plurality of the Court reasoned that the exemption lacked "sufficient breadth to pass scrutiny under the Establishment Clause. Every tax exemption constitutes a subsidy that affects nonqualifying taxpayers, forcing them to become 'indirect and vicarious "donors."' . . . [W]hen government directs a subsidy exclusively to religious organizations that is not required by the Free Exercise Clause and that either *burdens nonbeneficiaries markedly* or cannot reasonably be seen as removing a significant state-imposed deterrent to the free exercise of religion, it 'provide[s] unjustifiable awards of assistance to religious organizations' and cannot but 'conve[y] a message of endorsement' to slighted members of the community." Id. at 14–15 (emphasis added).

The strongest argument against the doctrine of required religious accommodations is that accommodations only for religious practices are impermissible under the nonestablishment norm. Nonetheless, even as a plurality of the Court in *Smith* rejected the position that under the free exercise norm government sometimes *must* accommodate a religious practice by exempting it, it reaffirmed the position that under the nonestablishment norm government sometimes *may* accommodate a religious practice by exempting it. See 494 U.S. at 890 (1990). For a helpful articulation of criteria for the "permissible" accommodation of religious practice, see McConnell, "Accommodation of Religion," n. 3, at 98–708.

79. Although some Buddhist sects are theistic, Buddhism—unlike Christianity, for example—is predominantly nontheistic, in the sense that Buddhism does not affirm the meaningfulness of "God"-talk. Nonetheless, Buddhism does seem to affirm the existence of a transcendent reality that is the source, the ground, and the end of everything else. See John B. Cobb, Jr. & Christopher Ives, eds., The Emptying God: A Buddhist-Jewish-Christian Conversation (1990); see, in particular, David Tracy, "Kenosis, Sunyata, and Trinity: A Dialogue With Masao Abe," id., at 135.

80. My position, in this chapter, about the constitutionally permissible role of religion in politics—like my position in the next two chapters about the morally proper role of religion in politics—is meant to apply to atheistic arguments as well as to religious ones: arguments that presuppose the truth of and include as one of their essential elements the belief that God does not exist. In a society that, like the United States, is overwhelmingly religious, it would not be acceptable to deprivilege religious arguments relative to atheistic ones. As Kent Greenawalt has cautioned, "[O]ne must present grounds for the [proposed principle of restraint] that have appeal to persons of religious and ethical views different from one's own." Kent Greenawalt, Private Consciences and Public Reasons 128 (1995). Cf. id. at 63 ("assum[ing] that a principle of restraint against reliance on religious grounds would also bar reliance on antireligious grounds").

81. By "other public official" I mean, here and elsewhere in this book, principally the policymaking officials in the executive branch of government. The chief policymaking official in the executive branch of the national government is, of course, the President of the United States; the chief policymaking official in the executive branch of a state government is the governor of the state. (On the judicial branch of government, see the appendix to ch. 3 of this book.)

82. See Esbeck, n. 33, at 604 n. 83. Cf. McDaniel v. Paty, 435 U.S. 618 (1978).

83. See Perry, The Constitution in the Courts, n. 6, at 143–49.

84. This is not to deny that as a constitutional matter government may require as a condition of continued employment that some of its employees (e.g., members of the police force) refrain from saying some things—at least, from saying some things openly—that they are constitutionally free to say (e.g., "Blacks aren't human"). See Pickering v. Board of Education, 391 U.S. 563 (1968).

85. See Douglas Laycock, "Freedom of Speech That Is Both Religious and Political," University of California at Davis L. Rev. (forthcoming, 1996). The Supreme Court recently remarked that "in Anglo-American history, at least, governmental suppression of speech has so commonly been directed *precisely* at religious speech that a free-speech clause without religion would be *Hamlet* with-

out the prince." Capitol Square Review and Advisory Board v. Pinette, 115 S.Ct. 2440, 2446 (1995).

86. See, e.g., Presbyterian Church (U.S.A.), God Alone is Lord of the Conscience: A Policy Statement Adopted by the 200th General Assembly 48 (1989): "[I]t is a limitation and denial of faith not to seek its expression in both a personal and a public manner, in such ways as will not only influence but transform the social order. Faith demands engagement in the secular order and involvement in the political realm."

87. David Hollenbach, SJ, "A Communitarian Reconstruction of Human Rights: Contributions from Catholic Tradition," in R. Bruce Douglass & David Hollenbach, SJ, eds., Catholicism and Liberalism 127, 142 (1994) (quoting Dignitatis Humanae, nos. 2 and 4).

88. Of course, presenting religious arguments in *nonpublic* political debate—political debate around the kitchen table, for example, or at a meeting of the local parish's Peace and Justice Committee—is not constitutionally problematic. A practical problem with the position that presenting religious arguments in public political debate is constitutionally problematic is that it might sometimes be difficult to say when "nonpublic" political debate has crossed the line and become "public". But that practical problem is also an academic one, because, as I have explained, presenting religious arguments in public politicial debate is not constitutionally problematic.

89. I explain, in ch. 2, why even one who opposes government basing political choices on religious arguments need not, and indeed should not, oppose legislators or other public officials, much less citizens, presenting religious arguments about the morality of human conduct in public political debate. See ch. 2, pp. 44–49.

90. Imagine this scenario: The Supreme Court invalidates a policy choice by one state (e.g., the state's refusal to recognize homosexual marriage) because a majority of the Court speculates that the state would not have made the choice but for a religious argument; two years later, the Court, with a slightly different membership, declines to invalidate the very same policy choice by another state because a majority of the Court speculates that the state would have made the choice even in the absence of a religious argument.

91. Sullivan, n. 49, at 197. See also Esbeck, n. 33, at 601–04.

92. Cf. Lawrence G. Sager, "Fair Measure: The Legal Status of Underenforced Constitutional Norms," 91 Harvard L. Rev. 1212 (1978).

93. Mark Tushnet has reached much the same conclusion by a different route. See Mark Tushnet, "The Limits of the Involvement of Religion in the Body Politic," in James E. Wood, Jr. & Derek Davis, eds., The Role of Religion in the Making of Public Policy 191 (1991).

Given the importance of the nonestablishment norm, and of the religious freedom it protects, once it has been established that a religious argument has played a nontrivial role in government making a political choice about the morality of human conduct, the party defending the choice in court properly bears the burden of final doubt about whether a plausible secular rationale supports the choice. Therefore, the defending party should be required to show that there is a plausible secular rationale, rather than the party challenging the choice required to show that no such rationale exists.

94. See Laycock, "Freedom of Speech That Is Both Religious and Political," n. 85, (commenting on a paper of mine).

95. "At the core of the Establishment Clause should be the principle that government cannot engage in a religious observance or compel or persuade citizens to do so." Id.

96. Indeed, one might want to argue that if government would not have made the coercive political choice but for a religious reason or reasons, then government has imposed religion even if a plausible secular rationale supports the choice. However, I have explained why as a practical matter it makes sense to accept "underenforcement" of this "ideal" understanding of what the nonestablishment norm forbids.

97. Cf. Federal Communications Commission v. Beach Communications, Inc., 113 S.Ct. 2096, 2101–03 (1993) (elaborating the "rational basis" test). However, those with the principal policymaking authority and responsibility—in particular, legislators—should ask themselves whether they find a secular rationale persuasive. See generally Paul Brest, "The Conscientious Legislator's Guide to Constitutional Interpretation," 27 Stanford L. Rev. 585 (1975).

98. See Perry, The Constitution in the Courts, n. 6, at 174–79; Baehr v. Lewin, 852 P.2d 44 (Hawaii 1993). Cf. Andrew Koppelman, "Why Discrimination against Lesbians and Gay Men Is Sex Discrimination," 69 New York University L. Rev. 197 (1994).

99. Cf. Sager, n. 92.

100. Even taking into account its unconstitutionality, an act might not be, all things considered, morally inappropriate. While relevant to an assessment of the morality of an act, that an act is unconstitutional does not by itself entail the immorality of the act. After all, one can imagine a constitution that forbids that which is morally required, or requires that which is morally forbidden.

101. Article V provides, in relevant part: "The Congress, whenever two thirds of both Houses shall deem it necessary, shall propose Amendments to this Constitution, or, on the Application of the Legislatures of two thirds of the several States, shall call a Convention for proposing Amendments, which, in either Case, shall be valid to all Intents and Purposes, as Part of this Constitution, when ratified by the Legislatures of three fourths of the several States, or by Conventions in three fourths thereof, as the one or the other Mode of Ratification may be proposed by the Congress[.]"

102. See n. 97.

103. For the texts of all these documents (and many others), see Brownlie, n. 64.

Chapter Two

1. Mark Tushnet, "The Limits of the Involvement of Religion in the Body Politic," in James E. Wood, Jr. & Derek Davis, eds., The Role of Religion in the Making of Public Policy 191, 213 (1991).

2. See ch. 3, pp. 82–85.

3. Luke Timothy Johnson, "Religious Rights and Christian Texts," in John Witte, Jr. & Johan David van der Vyver, eds., Religious Human Rights in Global Perspective: Religious Perspectives 65, 72–73 (1996).

4. Richard Rorty, "Religion as Conversation-Stopper," 3 Common Knowledge 1, 2 (1994).

5. Cf. Michael W. McConnell, "Political and Religious Disestablishment," 1986 Brigham Young U. L. Rev. 405, 413: "Religious differences in this country have never generated the civil discord experienced in political conflicts over such issues as the Vietnam War, racial segregation, the Red Scare, unionization, or slavery."

6. Cf. McDaniel v. Paty, 435 U.S. 618, 640–41 (1978) (Brennan, J. concurring in judgment):

> That public debate of religious ideas, like any other, may arouse emotion, may incite, may foment religious divisiveness and strife does not rob it of constitutional protection. . . . The mere fact that a purpose of the Establishment Clause is to reduce or eliminate religious divisiveness or strife, does not place religous discussion, association, or political participation in a status less preferred than rights of discussion, association and political participation generally. . . . The State's goal of preventing sectarian bickering and strife may not be accomplished by regulating religious speech and political association. . . . Government may not as a goal promote "safe thinking" with respect to religion . . . The Establishment Clause, properly understood, . . . may not be used as a sword to justify repression of religion or its adherents from any aspect of public life.

7. Cf. Daniel O. Conkle, "Different Religions, Different Politics: Evaluating the Role of Competing Religious Traditions in American Politics and Law," 10 J. L. & Religion 1 (1993–94).

8. To his credit, Richard Rorty insists that there is "hypocrisy . . . in saying that believers somehow have no right to base their political views on their religious faith, whereas we atheists have every right to base ours on Enlightenment philosophy. The claim that in doing so we are appealing to reason, whereas the religious are being irrational, is hokum." Rorty, n. 4, at 4.

9. David Tracy, Plurality and Ambiguity: Hermeneutics, Religion, Hope 84–85, 86, 97–98, 112 (1987).

10. As David Tracy has written, religion is "the single subject about which many intellectuals can feel free to be ignorant. Often abetted by the churches, they need not study religion, for 'everybody' already knows what religion is: It is a private consumer product that some people seem to need. Its former social role was poisonous. Its present privatization is harmless enough to wish it well from a civilized distance. Religion seems to be the sort of thing one likes 'if that's the sort of thing one likes.'" David Tracy, The Analogical Imagination 13 (1981). See also Kent Greenawalt, Religious Convictions and Political Choice 6 (1988):

"A good many professors and other intellectuals display a hostility or skeptical indifference to religion that amounts to a thinly disguised contempt for belief in any reality beyond that discoverable by scientific inquiry and ordinary human experience." Cf. "Special Issue—Religion and the Media: Three Forums," Commonweal, Feb. 24, 1995.

11. See Greenawalt, Religious Convictions and Political Choice, n. 10, at 159: "[I]f the worry is openmindedness and sensitivity to publicly accessible reasons, drawing a sharp distinction between religious convictions and [secular] personal bases [of judgment] would be an extremely crude tool."

David Tracy has lamented that "[f]or however often the word is bandied about, dialogue remains a rare phenomenon in anyone's experience. Dialogue demands the intellectual, moral, and, at the limit, religious ability to struggle to hear another and to respond. To respond critically, and even suspiciously when necessary, but to respond only in dialogical relationship to a real, not a projected other." David Tracy, Dialogue With the Other 4 (1990). Steven Smith, commenting wryly that "'dialogue' seems to have become the all-purpose elixir of our time", has suggested that "[t]he hard question is not whether people should talk, but rather what they should say and what (among the various ideas communicated) they should believe." Steven D. Smith, "The Pursuit of Pragmatism," 100 Yale L. J. 409, 434–35 (1990). As Tracy's observation suggests, however, there is yet another "hard" question, which Smith's suggestion tends to obscure. It is not *whether* but *how* people should talk; what qualities of character and mind should they bring, or try to bring, to the task.

12. David Hollenbach, SJ, "Civil Society: Beyond the Public-Private Dichotomy," 5 The Responsive Community 15, 22 (Winter 1994/95). One of the religious communities to which Hollenbach refers is the Catholic community. See David Hollenbach, SJ, "Contexts of the Political Role of Religion: Civil Society and Culture," 30 San Diego L. Rev. 877, 891 (1993):

> For example, the Catholic tradition provides some noteworthy evidence that discourse across the boundaries of diverse communities is both possible and potentially fruitful when it is pursued seriously. This tradition, in its better moments, has experienced considerable success in efforts to bridge the divisions that have separated it from other communities with other understandings of the good life. In the first and second centuries, the early Christian community moved from being a small Palestinian sect to active encounter with the Hellenistic and Roman worlds. In the fourth century, Augustine brought biblical faith into dialogue with Stoic and Neoplatonic thought. His efforts profoundly transformed both Christian and Graeco-Roman thought and practice. In the thirteenth century Thomas Aquinas once again transformed Western Christianity by appropriating ideas from Aristotle that he had learned from Arab Muslims and from Jews. In the process he also transformed Aristotelian ways of thinking in fundamental ways. Not the least important of these transformations was his insistence that

> the political life of a people is not the highest realization of the good of which they are capable—an insight that lies at the root of constitutional theories of limited government. And though the Church resisted the liberal discovery of modern freedoms through much of the modern period, liberalism has been transforming Catholicism once again through the last half of our own century. The memory of these events in social and intellectual history as well as the experience of the Catholic Church since the Second Vatican Council leads me to hope that communities holding different visions of the good life can get somewhere if they are willing to risk conversation and argument about these visions. Injecting such hope back into the public life of the United States would be a signal achievement. Today, it appears to be not only desirable but necessary.

See also id. at 892–896.

13. See Hollenbach, "Civil Society: Beyond the Public-Private Dichotomy," n. 12, at 22:

> Conversation and argument about the common good [including religious conversation and argument] will not occur initially in the legislature or in the political sphere (narrowly conceived as the domain in which conflict of interest and power are adjudicated). Rather it will develop freely in those components of civil society that are the primary bearers of cultural meaning and value—universities, religious communities, the world of the arts, and serious journalism. It can occur wherever thoughtful men and women bring their beliefs on the meaning of the good life into intelligent and critical encounter with understandings of this good held by other peoples with other traditions. In short, it occurs wherever education about and serious inquiry into the meaning of the good life takes place.

14. Hollenbach, "Contexts of the Political Role of Religion: Civil Society and Culture," n. 12, at 900. See also Kent Greenawalt, "Religious Convictions and Political Choice: Some Further Thoughts," 39 DePaul L. Rev. 1019, 1034 (1990) (expressing skepticism about "the promise of religious perspectives being transformed in what is primarily political debate").

15. Hollenbach, "Contexts of the Political Role of Religion: Civil Society and Culture," n. 12, at 900.

16. Cf. Paul G. Stern, "A Pluralistic Reading of the First Amendment and Its Relation to Public Discourse," 99 Yale L. J. 925, 934 (1990): "[W]e can freely and intelligently exercise our freedom of choice on fundamental matters having to do with our own individual ideals and conceptions of the good only if we have access to an unconstrained discussion in which the merits of competing moral, religious, aesthetic, and philosophical values are given a fair opportunity for hearing."

17. No one suggests that presenting religious arguments in *nonpublic* political debate—political debate around the kitchen table, for example, or at a

meeting of the local parish's Peace and Justice Committee—is morally problematic. A practical problem with the position that presenting religious arguments in public political debate is morally problematic is that it might sometimes be difficult to say when "nonpublic" political debate has crossed the line and become "public". Moreover, it is no more possible to maintain "an airtight barrier" between the religiously based moral discourse that takes place in nonpublic political debate and that which takes place in public political debate than it is to maintain an airtight barrier between the religiously based moral discourse that takes place in "universities, religious communities, the world of the arts, and serious journalism" and that which takes place in "the domains of government and policy-formation". Why not, then, just welcome the presentation of religiously based moral arguments in public as well as in relatively nonpublic political debate?

18. The following statement by Jürgen Habermas is noteworthy here:

> I do not believe that we, as Europeans, can seriously understand concepts like morality and ethical life, person and individuality, or freedom and emancipation, without appropriating the substance of the Judeo-Christian understanding of history in terms of salvation. And these concepts are, perhaps, nearer to our hearts than the conceptual resources of Platonic thought, centering on order and revolving around the cathartic intuition of ideas. Others begin from other traditions to find the way to the plenitude of meaning involved in concepts such as these, which structure our self-understanding. But without the transmission through socialization and the transformation through philosophy of *any one* of the great world religions, this semantic potential could one day become inaccessible. If the remnant of the intersubjectively shared self-understanding that makes human(e) intercourse with one another possible is not to disintegrate, this potential must be mastered anew by every generation.

Jürgen Habermas, Postmetaphysical Thinking: Philosophical Essays 15 (1992).

19. John A. Coleman, SJ, An American Strategic Theology 192–95 (1982). Coleman adds: "I am further strongly convinced that the Enlightenment desire for an unmediated universal fraternity and language (resting as it did on unreflected allegiance to *very particular* communities and language, conditioned by time and culture) was destructive of the lesser, real 'fraternities'—in [Wilson Carey] McWilliams' sense—in American life." Id. at 194.

Cf. John A. Coleman, SJ, "A Possible Role for Biblical Religion in Public Life," in David Hollenbach, SJ, ed., "Theology and Philosophy in Public: A Symposium on John Courtney's Murray's Unfinished Agenda," 40 Theological Studies 701, 704 (1979):

> American Catholic social thought in general and [John Courtney] Murray in particular appealed generously to the American liberal tradition of public philosophy and the classic understanding of republican virtue embedded in the medieval synthesis. Curiously, however, they

were very sparing in invoking biblical religion and the prophetic tradition in their efforts to address issues of public policy.

There are two reasons for this Catholic reluctance to evoke biblical imagery in public discourse. Much of the public religious rhetoric for American self-understanding was couched in a particularist Protestant form which excluded a more generously pluralistic understanding of America. Perhaps one reason why American Catholics and Jews have never conceived of the American proposition as a covenant—even a broken one—was because Protestant covenant thought tended in practice to exclude the new immigrants. Hence, for American Catholics as for Jews, more "secular" Enlightenment forms and traditions promised inclusion and legitimacy in ways Protestant evangelical imagery foreclosed. As Murray states it, the Protestant identification with America led to "Nativism in all its manifold forms, ugly and refined, popular and academic, fanatic and liberal. The neo-Nativist as well as the paleo-Nativist addresses to the Catholic substantially the same charge: 'You are among us but not of us.'" . . . [Murray] made no religious claims for the founding act of America as such. Catholics, decidedly, were not here in force when the Puritans and their God made a covenant with the land. Nor were they ever conspicuously invited to join the covenant. They preferred, therefore, a less religious, more civil understanding of America.

The second reason for a Catholic predilection for the two traditions of republican theory and liberal philosophy is the Catholic recognition of the need for secular warrant for social claims in a pluralist society. This penchant is rooted in Catholic natural-law thought.

20. Jeremy Waldron, "Religious Contributions in Public Deliberation," 30 San Diego L. Rev. 817, 841–42 (1993). Cf. Michael J. Sandel, "Political Liberalism," 107 Harvard L. Rev. 1765, 1794 (1994): "It is always possible that learning more about a moral or religious doctrine will lead us to like it less. But the respect of deliberation and engagement affords a more spacious public reason than liberalism allows. It is also a more suitable ideal for a pluralist society. To the extent that our moral and religious disagreements reflect the ultimate plurality of human goods, a deliberative mode of respect will better enable us to appreciate the distinctive goods our different lives express."

21. See n. 4 and accompanying text.

22. As I have indicated in chapter 1, I am in substantial agreement with the position that, in Kathleen Sullivan's formulation, "the negative bar against establishment of religion implies the affirmative 'establishment' of a civil order for the resolution of public moral disputes. . . . [P]ublic moral disputes may be resolved only on grounds articulable in secular terms." Kathleen M. Sullivan, "Religion and Liberal Democracy," 59 U. Chicago L. Rev. 195, 197 (1992). However, Sullivan is wrong to suggest that the fact that government may not make political choices in the absence of a plausible secular rationale constitutes "the banishment of religion from the public square . . ." Id. at 222. First, "the public

square"—the public culture of a society—includes much more than politics. To banish religion from politics is not to banish it from the rest of public culture. Second, religion has not been banished even from politics (much less from the rest of public culture): As I have explained, it is neither constitutionally nor morally inappropriate for legislators or other public officials, much less citizens, to present religiously based arguments about the morality of human conduct in public political debate. Indeed, because of the role that such religious arguments inevitably play in the political process, it is important that such arguments, no less than secular moral arguments, be presented in—so that they can be tested in—public political debate.

23. Greenawalt defends his position in his recent book. See Kent Greenawalt, Private Consciences and Public Reasons (1995). Greenawalt emphasizes that his position on the morally proper role of religion in politics is designed not for every liberal democratic society but for a particular one: the United States. Greenawalt understands that the precise arrangement between religion and politics that makes the most sense for the United States, given its history and traditions, political culture, and present circumstances, might not make the most sense for another liberal democratic society with a relevantly different history, political culture, etc. Moreover, Greenawalt's principal aim is not to recommend an arrangement that the law should impose, but only principles of *self*-restraint: principles—an arrangement—that citizens and legislators and other public officials should impose on themselves. Put another way, Greenawalt's aim is to recommend an informal (nonlegal) arrangement for the United States; it is to recommend that a certain understanding, that certain expectations, be established (or, if already established, maintained) in American political culture. Greenawalt understands that even if the law does not and should not forbid (or require) an activity, there might nonetheless be good reasons for persons not to engage (or to engage) in the activity.

There is much in Greenawalt's thoughtful book with which I am in substantial agreement and on which I do not comment here. In particular, I concur in most of Greenawalt's balanced but critical comments on the positions of several other contributors to the debate about religion in politics.

24. Id. at 156.

25. Id. at 157.

26. Id. at 132.

27. Steven D. Smith, Foreordained Failure: The Quest for a Constitutional Principle of Religious Freedom 164–65 n. 66 (1995). Smith quotes Mark Tushnet: "[N]onadherents who believe that they are excluded from the political community are merely expressing the disappointment felt by everyone who has lost a fair fight in the arena of politics." Mark V. Tushnet, "The Constitution of Religion," 18 Connecticut L. Rev. 701, 712 (1986) (quoted in Smith, this note, at 164–65 n. 66).

28. Greenawalt, Private Consciences and Public Reasons, n. 23, at 157.

29. This is the antidiscrimination language of Article 2 of the Universal Declaration of Human Rights.

30. Greenawalt, Private Consciences and Public Reasons, n. 23, at 157.

31. Of course, it might happen to be the case, in a particular situation, that the best way to aim at the good of all is to attend to the needs, and therefore to the good, of some: Perhaps in a particular situation satisfying the needs of some is instrumentally related to achieving the good of all; or perhaps the needs of some are especially severe or have too long been neglected.

32. Cf. Edmund Burke, "Speech to the Electors of Bristol at the Conclusion of the Poll, Nov. 3, 1774," in Edmund Burke on Government, Politics, and Society 157 (B. Hill, ed., 1976): "Your representative owes you, not his industry only, but his judgment; and he betrays, instead of serving you, if he sacrifices it to your opinion."

33. Greenawalt, Private Consciences and Public Reasons, n. 23, at 158.

34. Id. at 157. See also id. at 130–31: "Although religious violence is now rare [in the United States], we are not yet close to a state of bland tolerance."

35. See Maimon Schwarzschild, "Religion and Public Debate in a Liberal Society: Always Oil and Water or Sometimes More Like Rum and Coca-Cola?," 30 San Diego L. Rev. 903 (1993).

36. Greenawalt, Private Consciences and Public Reasons, n. 23, at 166.

37. Lawrence B. Solum, "Faith and Justice," 39 DePaul U. L. Rev. 1083, 1096 (1990). Solum is stating the argument, not making it. Indeed, Solum is wary of the argument. See id. at 1096–97. Solum cites, as an instance of the argument, Stephen L. Carter, "The Religiously Devout Judge," 64 Notre Dame L. Rev. 932, 939 (1989). For another instance, see Maimon Schwarzschild, "Religion and Public Debate in a Liberal Society: Always Oil and Water or Sometimes More Like Rum and Coca-Cola?," 30 San Diego L. Rev. 903 (1993).

38. See Greenawalt, Private Consciences and Public Reasons, n. 23, at 152.

39. Id. at 158.

40. See id. at 157.

41. In correspondence, Gerry Whyte of the Trinity College (Dublin) Faculty of Law has written:

> Taking the example of abortion, if I oppose abortion for religious reasons . . . , then unless I can publicly declare those reasons, I will not be able to defend my stance from allegations that I am, say, misogynist. (Divorce, in the Irish context, would probably be another good example of this point.) In other words, even where my religious beliefs cannot persuade my interlocutor to change his/her views, still I must be allowed to cite them if only to establish the bona fide nature of my motives. I think that this is important because one can still respect the sincerity of the fundamentalist (and pay his/her argument certain dues as a result) even if at the end of the day the argument is not persuasive; however one would have no time at all for the person whose motives are known to be less than honest.

Letter to Michael J. Perry, July 12, 1994.

If the better view is that legislators—who, after all, represent many citizens other than just themselves—may present religious arguments in public political debate, *a fortiori* citizens—who represent only themselves—may present such arguments. (To say that citizens represent just themselves is not to say that citizens, in making political choices, should aim to secure merely what they believe to be good for them; as an ideal matter, they should aim to secure what they believe to be the common good. See Michael J. Perry, The Constitution in the Courts: Law or Politics? 104–05 (1994). Of course, they might understandably believe—and in some cases they might even quite plausibly believe—that what is good for them, or good for them especially, is also in the common good.) Greenawalt, whose view is that legislators should not present religious arguments in public political debate, is more permissive for ordinary citizens than for legislators. He writes that in making arguments in support of political choices in the small settings in which they typically express themselves, ordinary citizens may rely, *inter alia*, on their religious beliefs. "Most citizens never get involved in advocacy of political positions, beyond talking to family, close friends, and associates. In those personal settings, people should feel free to express their reliance on any grounds they find compelling . . ." Greenawalt, Private Consciences and Public Reasons, n. 23, at 160.

That Greenawalt's position is more permissive for ordinary citizens than for legislators is an instance of an instinct on the part of many persons to think that, with respect to making political choices or making public arguments in support of political choices or both, what is legitimate for citizens to do, because they represent only themselves, is not necessarily legitimate for legislators to do, because they represent many citizens. I agree with Jeremy Waldron, however, that, with respect to the matters under discussion here, it is a mistake to distinguish between what citizens may do and what their representatives may do. At least, it is a mistake to draw the distinction too sharply, or to put much weight on the distinction. See Waldron, n. 20, at 827–31. See also Tushnet, "The Limits of the Involvement of Religion in the Body Politic," n. 1, at 199–201 (arguing that it does not make sense to distinguish between the grounds on which citizens may rely, in making political choices, and the grounds on which their elected representatives may rely).

Although Greenawalt's position is permissive for ordinary citizens, it is not permissive for those whom Greenawalt calls "quasi-public citizens": media commentators, newspaper editors, presidents of large corporations, etc. According to Greenawalt, quasi-public citizens, like legislators, should avoid presenting religious arguments in public political debate. See Greenawalt, Private Consciences and Public Reasons, n. 23, at 160–61. ("One class of quasi-public citizens falls outside my conclusions here: those whose profession (chosen by others or self-chosen) is to speak from religious and other comprehensive perspectives." Id. at 161.) I have indicated why I think Greenawalt is wrong to ask legislators to forgo presenting religious arguments in public political debate. If I am right about that, a fortiori Greenawalt is wrong in asking quasi-public citizens to forgo presenting religious arguments in public political debate. This is not to deny that such citizens, like legislators, often have strategic reasons for featuring, in their public political advocacy, secular reasons. But that legislators and others often have strategic reasons for downplaying religious reasons does not mean that they should forgo presenting religious arguments in public political debate if they want to do so. (In addition to citizens and legislators, Greenawalt discusses judges. I comment on that aspect of Greenawalt's position in the appendix to ch. 3.)

42. See John Rawls, Political Liberalism 222–54 (1993). See also Lawrence B. Solum, "Constructing an Ideal of Public Reason," 30 San Diego L. Rev. 729 (1993).

In his new introduction to the paper edition (1996) of *Political Liberalism*, Rawls has revised his earlier discussion of the idea of public reason. In particular, Rawls writes (at pp. li–lii of the paper edition): "When engaged in public reasoning may we also include reasons of our comprehensive doctrines? I now believe . . . that reasonable such doctrines may be introduced in public reason at any time, provided that in due course public reasons, given by a reasonable political conception, are presented sufficient to support whatever the comprehensive doctrines are introduced to support. I refer to this as the provision and it specifies what I now call the wide view of public reason." Then, in a footnote (at p. lii), Rawls says: "[M]any questions need to be considered in applying the proviso. One is, when does it need to be satisfied, on the same day or on some later day? Also, on whom does the obligation to honor it fall? There are many such questions, I only indicate a few of them here. The point is that it ought to be clear and established how the proviso is to be appropriately satisfied."

43. See, e.g., Rawls, Political Liberalism, n. 42, at 215 (emphasis added).

44. Rawls's principal aim, however, is not to recommend a regime that the law should impose, but only a regime that citizens and public officials should impose on themselves. As with Kent Greenawalt, Rawls's aim is to recommend an informal arrangement for the United States and for relevantly similar societies; it is to recommend that a certain understanding, that certain expectations, be established in the political culture of such a society.

45. Rawls, Political Liberalism, n. 42, at 220. Rawls adds: "These reasons are social and certainly not private." Then, in a footnote, he says: "The public v. nonpublic distinction is not the distinction between public and private. The latter I ignore: there is no such thing as private reason." Id. at 220 & n. 7.

46. Id. at 214.

47. Id. at 215. Rawls then adds: "Yet this may not always be so."

48. Solum, "Constructing an Ideal of Public Reason," n. 42, at 738–39. Cf. Greenawalt, Private Consciences and Public Reasons, n. 23, at 117–20 (arguing that Rawls's two-track approach—public reason for political questions that concern "the most fundamental matters" but not for other political questions—gives rise to "[s]erious technical problems [that raise] further doubt about the theoretical defensibility of his position and about the feasibility of its practical application[ ]").

49. I comment on Rawls's position with respect to judges in the appendix to ch. 3.

50. See n. 51.

51. See Rawls, Political Liberalism, n. 42, at 215–16:

> Another feature of public reason is that its limits do not apply to our personal deliberations and reflections about political questions, or to the reasoning about them by members of associations such as churches and universities, all of which is a vital part of the background culture. Plainly, religious, philosophical, and moral considerations of many kinds may here properly play a role. But the ideal of public reason does hold for citizens when they engage in political advocacy in the public forum, and thus for members of political parties and for candidates in their campaigns and for other groups who support them. It holds equally for how citizens are to vote in elections when constitutional essentials or matters of basic justice are at stake. Thus, the ideal of public reason not only governs the public discourse of elections insofar as the issues involve those fundamental questions, but also how citizens are to cast their vote on these questions. Otherwise, public discourse runs the risk of being hypocritical: citizens talk before one another one way and vote another. . . . [T]he ideal of public reason . . . applies in official forums and so to legislators when they speak on the floor of parliament, and to the executive in its public acts and pronouncements.

Rawls adds that the ideal of public reason "applies . . . in a special way to the judiciary and above all to a supreme court in a constitutional democracy with judicial review." Id. at 216. Again, I comment on Rawls's position with respect to judges in the appendix to ch. 3.

52. Id. at 225.

53. Id. at 224.

54. Id. at 213. See also id. at 169 ("political values expressed by the political conception [of justice] endorsed by the overlapping consensus").

55. Id. at 224. "Now it is essential that a liberal political conception include, besides its principles of justice, guidelines of inquiry that specify ways of reasoning and criteria for the kinds of information relevant for political questions. Without such guidelines substantive principles cannot be applied and this leaves the political conception incomplete and fragmentary." Id. at 223–24.

56. Id. at 224.

57. Id. at 226–27.

58. In the new introduction to the paper edition of *Political Liberalism*, n. 42, Rawls writes (at p. xlvi): "[O]ur exercise of political power is proper only when citizens sincerely believe that the reasons offered for their political actions are not only sufficient but they *reasonably* think that other citizens might also *reasonably* accept those reasons." (Emphasis added.)

59. Rawls, Political Liberalism, n. 42, 133 et seq.

60. See Michael J. Perry, Love and Power: The Role of Religion and Morality in American Politics, ch. 6 (1991).

61. See Lawrence B. Solum, "On the Indeterminacy Crisis: Critiquing Critical Dogma," 54 U. Chicago L. Rev. 462 (1987).

62. In the new introduction to the paper edition of *Political Liberalism*, n. 42, Rawls writes (at p. liii): "One objection to the wide view of public reason is that it is still too restrictive. However, to establish this, we must find pressing questions of constitutional essentials or matters of basic justice that cannot reasonably be resolved by political values expressed by any of the existing reasonable political conceptions, nor also by any such conceptions that could be worked out. PL doesn't argue that this can never happen; it only suggests that it rarely does so. Whether public reason can settle all, or almost all, political questions by a reasonable ordering of political values cannot be decided in the abstract independent of actual cases."

63. I have discussed this state of affairs—and how the judiciary should respond to it—elsewhere. See generally Perry, The Constitution in the Courts, n. 41, esp. chs. 5–6.

64. See Philip Quinn, "Political Liberalisms and Their Exclusion of the Religious," 69 Proceedings & Addresses of the American Philosophical Association 35 (1995). I comment below on Rawls's characterization of the abortion controversy. See n. 74.

65. Rawls, Political Liberalism, n. 42, at 216.

66. Id. at 217. Rawls adds: "This duty also involves a willingness to listen to others and a fairmindedness in deciding when accommodations to their views should reasonably be made."

67. See ch. 3, paragraph accompanying n. 6. Commenting critically on Rawls's *Political Liberalism*, Michael Sandel has observed that "[o]n the liberal conception [of mutual respect], we respect our fellow citizens' moral and religious convictions by ignoring them (for political purposes), by leaving them undisturbed, or by carrying on political debate without reference to them. To admit moral and religious ideals into political debate about justice would undermine mutual respect in this sense." Sandel then remarks that "this is not the only, or perhaps even the most plausible way of understanding the mutual respect on which democratic citizenship depends. On a different conception of respect . . . we respect our fellow citizens' moral and religious convictions by engaging them, or attending to them—sometimes by challenging and contesting them, sometimes by listening and learning from them—especially if those convictions bear on important political questions." Sandel, n. 20, at 1794.

68. Gerald R. Dworkin, "Equal Respect and the Enforcement of Morality," 7 Social Philosophy & Policy 180, 193 (1990).

69. See Timothy P. Jackson, "Love in a Liberal Society: A Response to Paul J. Weithman," 22 J. Religious Ethics 29 (1994).

70. John Langan, SJ, "Overcoming the Divisiveness of Religion: A Response to Paul J. Weithman," 22 J. Religious Ethics 47, 51 (1994). Cf. Paul F. Campos, "Secular Fundamentalism," 94 Columbia L. Rev. 1814 (1994).

71. See Sandel, n. 20, at 1776, 1777: "Where grave moral questions are concerned, whether it is reasonable to bracket moral and religious controversies for the sake of political agreement partly depends on which of the contending moral or religious doctrines is true. . . . Even granting the importance of secur-

ing social cooperation on the basis of mutual respect, what is to guarantee that this interest is always so important as to outweigh any competing interest that could arise from within a comprehensive moral or religious view?"

72. See, e.g., pp. lv–lvii of the new introduction to the paper edition of *Political Liberalism*, n. 42.

73. See ch. 3, n. 34 and accompanying text.

74. See Quinn, n. 64. A default position in favor of a noncoercive public policy is deeply problematic. See Perry, Love and Power, n. 60, at 13–14.

In the new introduction to the paper edition of *Political Liberalism*, n. 42, Rawls seems to think that the abortion controversy sometimes exemplifies (whatever else it exemplifies) not the problem of the underdeterminacy of public reason but rather a "stand-off" between two competing arguments, one of them "pro-life", and the other "pro-choice", each of which is consistent with public reason. See p. lv:

> However, disputed questions, such as that of abortion, may lead to a stand-off between different political conceptions and citizens must simply vote on the question. Indeed, this is the normal case, unanimity of views is not to be expected. Reasonable political conceptions do not always lead to the same conclusion; nor do citizens holding the same conception always agree on particular issues. Yet the outcome of the vote is to be seen as reasonable provided all citizens of a reasonably just constitutional regime sincerely vote in accordance with the idea of public reason. This doesn't mean the outcome is true or correct, but it is for the moment reasonable, and binding on citizens by the majority principle. Some may, of course, reject a decision, as Catholics may reject a decision to grant a right to abortion. They may present an argument in public reason for denying it and fail to win a majority. But they need not exercise the right of abortion in their own case. They can recognize the right as belonging to legitimate law and therefore do not resist it with force.

As I have suggested in the text accompanying this note, however, the problem is not that (which Rawls calls "the normal case") of a "stand-off" between two arguments, each of which is consistent with the ideal of public reason. The problem, rather, is that of the underdeterminacy of public reason with respect to the policy issue of abortion.

Chapter Three

1. See ch. 1, n. 97.

2. See ch. 2, n. 66 and accompanying text.

3. William A. Galston, Liberal Purposes 108–09 (1991).

4. Michael J. Perry, "Religious Morality and Political Choice: Further Thoughts—and Second Thoughts—on *Love and Power*," 30 San Diego L. Rev. 703, 711 n. 23 (1993) (quoting Larmore).

5. Robert Audi, "The Place of Religious Argument in a Free and Democratic Society," 30 San Diego L. Rev. 677, 701 (1993).

6. Gerald R. Dworkin, "Equal Respect and the Enforcement of Morality," 7 Social Philosophy & Policy 180, 193 (1990) (criticizing Ronald Dworkin). See also John M. Finnis, Natural Law and Natural Rights 221–22 (1980) (criticizing Ronald Dworkin).

7. See ch. 2, n. 37.

8. John Courtney Murray, We Hold These Truths 23–24 (1960).

9. Kent Greenawalt, Private Consciences and Public Reasons 130 (1995).

10. See pp. 79–80.

11. That all human beings are sacred is represented in the Universal Declaration of Human Rights (Article 2) by this language: "Everyone is entitled to all the rights and freedoms set forth in this Declaration, without distinction of any kind, such as race, colour, sex, language, religion, political or other opinion, national or social origin, property, birth or other status."

A political choice based on the view that only white persons are sacred would violate the freedom from racial discrimination protected by the constitutional law of the United States. See Michael J. Perry, The Constitution in the Courts: Law or Politics? 143 et seq. (1994).

12. See n. 34 and accompanying text.

13. Hilary Putnam, The Many Faces of Realism 60–61 (1987).

14. Florida Bishops, "Promoting Meaningful Welfare Reform," 24 Origins 609, 611 (1995) (quoting Matthew 25:40). There are many such examples. See, e.g., National Conference of Catholic Bishops, Economic Justice for All: A Pastoral Letter on Catholic Social Teaching and the U. S. Economy v (1986): "This letter is a personal invitation to Catholics to use the resources of our faith, the strength of our economy, and the opportunities of our democracy to shape a society that better protects the dignity and basic rights of *our sisters and brothers both in this land and around the world*." (Emphasis added.) By "our sisters and brothers", the Catholic bishops meant, not "our fellow Catholics" or even "our fellow Christians", but "all human beings".

15. I have discussed the idea of human rights at length elsewhere. See Michael J. Perry, The Idea of Human Rights: Four Inquiries (forthcoming, 1998).

16. For the texts of all the human rights documents to which I refer in this paragraph, see Ian Brownlie, ed., Basic Documents on Human Rights (1992).

17. The representatives of 172 states adopted by consensus The Vienna Declaration and Programme of Action.

18. Article 2 continues: "Furthermore, no distinction shall be made on the basis of the political, jurisdictional or international status of the country or territory to which a person belongs, whether it be independent, trust, non-self-governing or under any other limitation of sovereignty."

19. In an essay on "The Spirituality of The Talmud", Ben Zion Bokser and Baruch M. Bokser write: "From this conception of man's place in the universe comes the sense of the supreme sanctity of all human life. 'He who destroys one person has dealt a blow at the entire universe, and he who sustains or saves one

person has sustained the whole world.'" Ben Zion Bokser & Baruch M. Bokser, "Introduction: The Spirituality of the Talmud," in The Talmud: Selected Writings 7 (1989). They continue:

> The sanctity of life is not a function of national origin, religious affiliation, or social status. In the sight of God, the humble citizen is the equal of the person who occupies the highest office. As one talmudist put it: "Heaven and earth I call to witness, whether it be an Israelite or pagan, man or woman, slave or maidservant, according to the work of every human being doth the Holy Spirit rest upon him." . . . As the rabbis put it: "We are obligated to feed non-Jews residing among us even as we feed Jews; we are obligated to visit their sick even as we visit the Jewish sick; we are obligated to attend to the burial of their dead even as we attend to the burial of the Jewish dead."

Id. at 30–31.

20. Ronald Dworkin, "Life is Sacred. That's the Easy Part," New York Times Magazine, May 16, 1993, at 36. Cf. Ronald Dworkin, Life's Dominion: An Argument about Abortion, Euthanasia, and Individual Freedom 25 (1993) ("sacred" is often used in a theistic sense, but it can be used in a secular sense as well). (I have criticized Dworkin's conception of "sacred". See Michael J. Perry, "The Gospel According to Dworkin," 11 Constitutional Commentary 163 (1993).)

21. James Griffin, Well-Being: Its Meaning, Measurement, and Moral Importance 239 (1987).

22. Bernard Williams, Ethics and the Limits of Philosophy 14 (1985).

23. See Michael J. Perry, "Is the Idea of Human Rights Ineliminably Religious?," in Austin Sarat & Thomas R. Kearns, eds., Legal Rights: Historical and Philosophical Perspectives 205 (1996). (A later version of this essay will appear as ch. 1 of Perry, The Idea of Human Rights, n. 15.)

24. See n. 11.

25. See Dan Cohn-Sherbok, ed., World Religions and Human Liberation (1992); Hans Küng & Jürgen Moltmann, eds., The Ethics of World Religions and Human Rights (1990); Leroy S. Rouner, ed., Human Rights and the World's Religions (1988); Arlene Swidler, ed., Human Rights in Religious Traditions (1982); Robert Traer, Faith in Human Rights: Support in Religious Traditions for a Global Struggle (1991).

26. And putting aside imaginable but utterly fantastic scenarios.

27. See Elizabeth Mensch & Alan Freeman, The Politics of Virtue: Is Abortion Debatable? (1993).

28. The Constitution of the United States has been interpreted by the Supreme Court to ban legislation outlawing abortion in the period of pregnany prior to viability. The principal case is *Roe v. Wade*, 410 U.S. 113 (1973), which was substantially reaffirmed in *Planned Parenthood of Southeastern Pennsylvania v. Casey*, 505 U.S. 833 (1992). See Perry, The Constitution in the Courts, n. 11, at 179–89 (1994).

The Constitution of the Federal Republic of Germany, by contrast, has been interpreted by the German Supreme Court to ban legislation permitting most abortions in the first three months of pregnancy. In 1993, the Constitutional Court of the Federal Republic of Germany ruled that Germany's new liberal abortion law "was unconstitutional because it violates a constitutional provision requiring the state to protect human life." Stefan Kinzer, "German Court Restricts Abortion, Angering Feminists and the East," New York Times, May 29, 1993, at A1. See Judgment of May 28, 1993, The Constitutional Court of the Federal Republic of Germany, Judgment of the Second Senate, 20 EeGRZ 229–275 (consolidated case nos. 2 BxF2/90m 2 BzF 4/92, 2 BzF 5/92).

The Constitution of Ireland, in Article 40.3.3, states: "The State acknowledges the right to life of the unborn and, with due regard to the equal right to life of the mother, guarantees in its laws to respect, and, as far as practicable, by its laws to defend and vindicate that right." Article 40.3.3 also provides: "This subsection shall not limit freedom to travel between the State and another state. This subsection shall not limit freedom to obtain or make available, in the State, subject to such conditions as may be laid down by law, information relating to services lawfully available in another state." On Article 40.3.3 and the abortion controversy in Ireland, see Gerard Hogan & G.F. Whyte, J.M. Kelly's The Irish Constitution 790–810 (1994). See also Jeffrey A. Weinstein, "'An Irish Solution to an Irish Problem': Ireland's Struggle With Abortion Law," 10 Arizona J. International & Comparative L. 165 (1993).

The debate about what public policy regarding abortion should be is closed off neither by judicial decisions like the two just referred to—decisions interpreting the constitution to dictate a particular political choice or range of political choices regarding abortion—nor even by a constitutional provision like Ireland's: Such decisions and such a provision leave open the question whether to amend the constitution to allow for, or to dictate, a different choice or range of choices.

29. It is not the official Catholic position that infliction of the death penalty is always immoral, although Pope John Paul II has declared in a recent encyclical that the conditions under which infliction of the death penalty is morally justified "are very rare if not practically nonexistent." John Paul II, "Evangelium Vitae," 24 Origins 689, 709 (1995).

30. Here is a sampling of the Catholic bishops' statements on abortion:

> [A]bortion . . . negates two of our most fundamental moral imperatives: respect for innocent life and preferential concern for the weak and defenseless. . . . "Because victims of abortion are the most vulnerable and defenseless members of the human family, it is imperative that we, as Christians called to serve the least among us, give urgent attention and priority to this issue."

Catholic Bishops of the United States, "Resolution on Abortion," 19 Origins 395 (1989). See also Bishop James McHugh, "Political Responsibility and Respect for Life," 19 Origins 460 (1989).

> We would do well to pay special heed to the implications of the great commandment [to love our neighbor as ourselves, which requires] that we value the lives and needs of others no less than our own. The right to life of the unborn baby, of the ill and infirm grandparent, of the despicable criminal, of the AIDS patient, is to be affirmed and protected as though it belonged to us. In addition, the refugee from Indochina, the lives of the welfare recipient from Illinois and the homeless in our own community each possess a dignity that matches our own. When we respond to that need, we acknowledge not only their dignity but ours as well.

Catholic Bishops of Wisconsin, "A Consistent Ethic of Life," 19 Origins 461, 462 (1989).

> When people say abortion is a matter of choice, they're forgetting someone. "Pro-choice" is a phrase that is incomplete; it lacks an object. One must ask the natural follow-up: the choice to do what? If it were the choice to poison an elderly person, or to smuggle drugs, or to embezzle from a bank, no one would defend that choice. In this case, it's the choice to take [an unborn] child's life. Who defends . . . the child's inalienable right to life?

Archbishop John May, "Faith and Moral Teaching in a Democratic Nation," 19 Origins 385, 388 (1989).

> Catholic teaching sees in abortion a double moral failure: A human life is taken, and a society allows or supports the killing. Both concerns, protecting life and protecting the society from the consequences of destroying lives, require attention. Both fall within the scope of civil law. Civil law, of course, is not coextensive with the moral law, which is broader in its scope and concerns. But the two should not be separated; the civil law should be rooted in the moral law even if it should not try to translate all moral prohibitions and prescriptions into civil statutes.
>
> When should the civil law incorporate key moral concerns? When the issue at stake poses a threat to the public order of society. But at the very heart of public order is the protection of human life and basic human rights. A society which fails in either or both of these functions is rightfully judged morally defective.
>
> Neither the right to life nor other human rights can be protected in society without the civil law. . . . [O]ur objective, that the civil law recognize a basic obligation to protect human life, especially the lives of those [like unborn children] vulnerable to attack or mistreatment, is not new in our society. The credibility of civil law has always been tested by the range of rights it defends and the scope of the community it protects. To return to the analogy of civil rights: The struggle of the 1960s was precisely about extending the protection of the law to those unjustly deprived of protection.

Cardinal Joseph Bernardin, "The Consistent Ethic of Life after *Webster*," 19 Origins 741, 746 (1990). See also Bishop John Myers, "Obligations of Catholics and Rights of Unborn Children," 20 Origins 65, 68 (1990); Bishop John O'Connor, "Abortion: Questions and Answers," 20 Origins 97 (1990). Cf. Archbishop John Roach, "War of Words on Abortion," 20 Origins 88, 89 (1990) (responding to the accusation that a call for restrictive abortion legislation is a call to "legislate morality"): "That's not a new argument. In the heat of the civil rights debate, Martin Luther King Jr. was accused of wanting to legislate morality. He replied that the law could not make people love their neighbors, but it could stop their lynching them."

31. See David Smith, MSC, "What Is Christian Teaching on Abortion?," 42 Doctrine & Life 305, 316 (1992):

> As can be observed from this brief survey of certain Christian Churches, all agree that the human embryo has "value" and must be respected. The disagreement concerns what precisely is the "value" of the human embryo. One view, represented explicitly by the Roman Catholic Church, states that it has exactly the same value as any other human being. Another view, represented by a strong body of opinion in the Church of England, asserts that its value, prior to individuation (consciousness), is less than that of a human being in the proper sense of the word. A third view, represented by the Methodist Conference, would argue that its value depends on its stage of development: thus a progressively increasing value. The Baptist Union seems to favour a similar position, as does the Church in Wales and the Free Churches.

32. Laurence H. Tribe, "Will the Abortion Fight Ever End: A Nation Held Hostage," New York Times, July 2, 1990, at A13 (Op-Ed Page). For a passionate critique, by a pro-choice feminist, of some pro-choice feminists' explicit or implicit denial of the moral status of the fetus, see Naomi Wolf, "Our Bodies, Our Souls: Rethinking Pro-choice Rhetoric," New Republic, Oct. 16, 1995, at 26.

33. On such an argument, see this ch., pp. 79–80.

34. Robert F. George, Book Review, 88 American Political Science Rev. 445, 445–46 (1994) (reviewing Ronald Dworkin, Life's Dominion: An Argument About Abortion, Euthaniasia, and Individual Freedom (1993)). George adds: "Frankly, I doubt that this challenge can be met. In any event, Dworkin here fails to make much progress toward meeting it." Id. at 446. Cf. Mary Warnock, "The Limits of Toleration," in Susan Mendus & David Edwards, eds., On Toleration 123, 125(1987) (commenting on John Stuart Mill's failure to address, inter alia, the question "Who is to count as a possible object of harm?").

35. Cf. Judith Jarvis Thomson, "Abortion," Boston Review, Summer 1995, 11, 15: "If the legislature constrains the liberty [of a woman to have an abortion] on the ground of this doctrine [that the fetus has a right to life from the moment of conception], and declares that it is entitled to do so because *God says* the doc-

trine is true, then the legislature does violate the principle of separation of church and state. But no sensible contemporary opponent of abortion invites the legislature to do this. The opponent of abortion instead invites the legislature to constrain the liberty on the ground of this doctrine, and to declare that it is entitled to do so because the doctrine *is* true."

36. See n. 34 and accompanying text.

37. William Collinge has said, in correspondence, that "as a Catholic, I would say that *ultimate* human well-being is sharing in the life of God, participating as somehow befits our status as created beings in the divine Trinity. Talk of grace, 'beatific vision,' mystical union all points in the same direction. Many adherents of other religions have corresponding beliefs about what is ultimately best. How could there be a secular argument for something like that?" Letter to Michael J. Perry, Sept. 1, 1995.

38. See n. 129 and accompanying text. For a description of the religious mindset that yields such a religious argument, see James Davison Hunter, Culture Wars: The Struggle to Define America 120–21 (1991).

39. Letter to Michael J. Perry, Aug. 7, 1995.

40. Cf. Williams, n. 22, at 96: "[Preferred ethical categories] may be said to be given by divine command or revelation; in this form, if it is not combined with a grounding in human nature, the explanation will not lead us anywhere except into what Spinoza called 'the asylum of ignorance.'"

41. This is not to say either that the existence of a persuasive secular argument entails the persuasiveness of the religious argument or that the nonexistence of a persuasive secular argument entails that the religious argument is incorrect.

42. "In principle", because "[t]he participation by man in God's eternal law through knowledge . . . can be corrupted and depraved in such a way that the natural knowledge of good is darkened by passions and the habits of sin. For Aquinas, then, not all the conclusions of natural law are universally known, and the more one descends from the general to the particular, the more possible it is for reason to be unduly influenced by the emotions, or by customs, or by fallen nature." John Mahoney, SJ, The Making of Moral Theology: A Study of the Roman Catholic Tradition 105–06 (1987).

43. For an illuminating recounting, see id. at 103–15.

44. Id. at 106, 107, 109 (1987). Mahoney then adds: "[B]ut such human thinking is not always or invariably at its best." Id. at 109.

45. Basil Mitchell, "Should Law Be Christian?," Law & Justice, No. 96/97 (1988), at 12, 21.

46. See n. 48.

47. Audi, n. 5, at 699. One can accept Audi's point and nonetheless believe that there is no good secular argument for the foundational moral proposition that each and every human being is sacred. See n. 23.

48. Discussing usury, marriage, slavery, and religious freedom, John Noonan has demonstrated:

> Wide shifts in the teaching of moral duties, once presented as part of Christian doctrine by the magisterium, have occurred. In each case one can see the displacement of a principle or principles that had been taken as dispositive—in the case of usury, that a loan confers no right to profit; in the case of marriage, that all marriages are indissoluble; in the case of slavery, that war gives a right to enslave and that ownership of a slave gives title to the slave's offspring; in the case of religious liberty, that error has no rights and that fidelity to the Christian faith may be physically enforced. . . . In the course of this displacement of one set of principles, what was forbidden became lawful (the cases of usury and marriage); what was permissible became unlawful (the case of slavery); and what was required became forbidden (the persecution of heretics).

John T. Noonan Jr., "Development in Moral Doctrine," 54 Theological Studies 662, 669 (1993). See also Seán Fagan, SM, "Interpreting the Catechism," 44 Doctrine & Life 412, 416–17 (1994):

> A catechism is supposed to "explain", but this one does not say why Catholics have to take such a rigid, absolutist stand against artificial contraception because it is papal teaching, but there is no reference to the explicit centuries-long papal teaching that Jews and heretics go to hell unless they convert to the Catholic faith, or to Pope Leo X, who declared that the burning of heretics is in accord with the will of the Holy Spirit. Six different popes justified and authorised the use of slavery. Pius XI, in an encyclical at least as important as *Humane Vitae*, insisted that co-education is erroneous and pernicious, indeed against nature. The Catechism's presentation of natural law gives the impression that specific moral precepts can be read off from physical human nature, without any awareness of the fact that our very understanding of "nature" and what is "natural" can be coloured by our culture.

49. On self-critical rationality, see Michael J. Perry, Love and Power: The Role of Religion and Morality in American Politics, ch. 4 (1991).

50. Richard John Neuhaus, "Reason Public and Private: The Pannenberg Project," First Things, March 1992, at 55, 57.

51. True, one might well be more inclined to find persuasive a secular argument about the requirements of human well-being if one already accepts a religious argument that reaches the same conclusion. But there's nothing to be done about that.

52. See ch. 2, n. 59.

53. Edward Schillebeeckx, The Schillebeeckx Reader 263 (Robert Schreiter, ed., 1984).

54. See National Conference of Catholic Bishops, Challenge of Peace: God's Promise and Our Response (1983).

55. Bryan Hehir, "Responsibilities and Temptations of Power: A Catholic View," unpublished ms. (1988).

56. The Williamsburg Charter: A National Celebration and Reaffirmation of the First Amendment Religious Liberty Clauses 22 (1988).

57. Richard John Neuhaus, "Nihilism Without the Abyss: Law, Rights, and Transcendent Good," 5 J. L. & Religion 53, 62 (1987). In commenting on this passage, Stanley Hauerwas has said that "[r]ather than condemning the Moral Majority, Neuhaus seeks to help them enter the public debate by basing their appeals on principles that are accessible to the public." Stanley Hauerwas, "A Christian Critique of Christian America," in J. Roland Pennock & John W. Chapman, eds., Religion, Morality, and the Law 110, 118 (1988).

58. See Perry, Love and Power, n. 49, ch. 6.

59. I have discussed the value of ecumenical political dialogue elsewhere. See Perry, Love and Power, n. 49, ch. 6. See also David Lochhead, The Dialogical Imperative: A Christian Reflection on Interfaith Encounter 79 (1988): "In more biblical terms, the choice between monologue and dialogue is the choice between death and life. If to be human is to live in community with fellow human beings, then to alienate ourselves from community, in monologue, is to cut ourselves off from our own humanity. To choose monologue is to choose death. Dialogue is its own justification."

60. Robin W. Lovin, "Why the Church Needs the World: Faith, Realism, and the Public Life," unpublished ms. (1988 Sorenson Lecture, Yale Divinity School). Defending the moderate style of his participation in public discourse about abortion and other issues implicating what he has famously called "the consistent ethic of life", Cardinal Joseph Bernardin, Archbishop of Chicago, has said: "The substance of the consistent ethic yields a style of teaching it and witnessing to it. The style should . . . not [be] sectarian. . . . [W]e should resist the sectarian tendency to retreat into a closed circle, convinced of our truth and the impossibility of sharing it with others. . . . The style should be persuasive, not preachy. . . . We should be convinced we have much to learn from the world and much to teach it. A confident church will speak its mind, seek as a community to live its convictions, but leave space for others to speak to us, help us grow from their perspective . . . ." Cardinal Joseph Bernardin, "The Consistent Ethic of Life After *Webster*," 19 Origins 741, 748 (1990).

61. See n. 34 and accompanying text.

62. John Paul II, "Faith and Freedom: Text of the Homily Delivered at Mass in Baltimore," New York Times, Oct. 9, 1995, at B15. See also John Paul II, "Man Is Bound by Nature To Seek the Truth," L'osservatore Romano, December 1995 (Weekly ed.), at 1, 7:

> I am thinking here of the claim that a democratic society should delegate to the realm of private opinion its members' religious beliefs and the moral convictions which derive from faith. At first glance, this appears to be an attitude of necessary impartiality and "neutrality" on the part of society in relation to those of its members who follow different religious traditions or none at all. Indeed, it is widely held that this is the only enlightened approach possible in a modern pluralistic state.

> But if citizens are expected to leave aside their religious convictions when they take part in public life, does this not mean that society not only excludes the contribution of religion to its institutional life, but also promotes *a culture which redefines man as less than he is?* In particular, there are moral questions at the heart of every great public issue. Should critics whose moral judgments are informed by their religious beliefs be less welcome to express their most deeply held convictions? When that happens, is not democracy itself emptied of real meaning? Should not genuine pluralism imply that firmly held convictions can be expressed in vigorous and respectful dialogue?

63. James Tunstead Burtchaell, "The Sources of Conscience," 13 Notre Dame Mag. 20, 20–21 (Winter 1984–85). (On our neighbor always turning out to be the most unlikely person, see Luke 10:29–37 ("Parable of the Good Samaritan").) Burtchaell continues: "And when debate and dialogue and testimony do fructify into conviction, and conviction into consensus, nothing could be more absurd than to expect that consensus to be confined within a person's privacy or a church's walls. Convictions are what we live by. Do we have anything better to share with one another?" Burtchaell, this note, at 21. (For a revised version of Burtchaell's essay, and for several other illuminating essays by Father Burtchaell, see James Tunstead Burtchaell, The Giving and Taking of Life (1989).)

64. John A. Coleman, An American Strategic Theology 196 (1982).

65. Martin E. Marty, "When My Virtue Doesn't Match Your Virtue," 105 Christian Century 1094, 1096 (1988). Marty adds: "Of course, these communities and their spokespersons argue with one another. But so do philosophical rationalists." Id.

66. David Tracy, Plurality and Ambiguity: Hermeneutics, Religion, Hope 88 (1987).

67. Id. at 88–89.

68. Mitchell, n. 45, at 21.

Assume that one believes that God is, for all things, including all human beings, the beginning and the end, Alpha and the Omega. That God is, for all human beings, the Alpha and the Omega does not entail that the world God has created is one in which knowledge about the requirements of human well-being is not "in principle" achievable by those who do not believe in God. At the very least, it does not entail that the world is one in which knowledge *of the politically relevant sort* about the requirements of human well-being—knowledge of the sort relevant to the sorts of coercive political choices a liberal democracy committed to nonestablishment is free to make—is not achievable by those who do not believe in God (Buddhists, for example).

69. See, e.g., David W. Dunlap, "Some States Trying to Stop Gay Marriages Before They Start," New York Times, Mar. 15, 1995, at A10. For a discussion of the issue, see Richard D. Mohr, "The Case for Gay Marriage," 9 Notre Dame J. L., Ethics & Public Policy 215 (1995). Few people in the United States today

argue that criminal laws against homosexual sexual conduct should be enforced. Indeed, laws banning homosexual sexual conduct have become a concern of the international human rights movement. For example, one of the world's foremost nongovernmental human rights organizations, Amnesty International, has recently taken up the cause "not only [of] those arrested for advocating homosexual rights, but also [of] those arrested solely for homosexual acts or identity . . ." Amnesty International USA, "Breaking the Silence: Human Rights Violations Based on Sexual Orientation," Amnesty Action, Winter/Spring 1994, at 1. Two decisions of the European Court of Human Rights have invalidated laws banning homosexual sexual conduct. See Dudgeon v. United Kingdom, 45 Eur. Ct. H. R. (ser. A) (1981); Norris v. Ireland, 142 Eur. Ct. H. R. (ser. A) (1988). I have criticized the U.S. Supreme Court's failure, in *Bowers v. Hardwick*, 478 U.S. 176 (1986), to invalidate laws banning homosexual sexual conduct. See Perry, The Constitution in the Courts, n. 11, at 174–79.

Article 2 of the Universal Declaration of Human Rights, in language that is repeated both in the International Covenant on Economic, Social and Cultural Rights and in the International Covenant on Civil and Political Rights, explicitly forbids discrimination based on "race, colour, sex, language, religion, political or other opinion, national or social origin, property, birth or other status." It does not explicitly forbid discrimination based on sexual orientation. Recently, however, the Human Rights Committee of the United Nations, interpreting the International Covenant on Civil and Political Rights, has ruled that discrimination based on "sex" includes discrimination based on "sexual orientation". Nicholas Toonen v. Australia, Communication No. 488/1992 (Mar. 31, 1994). For an argument in support of such a construal, see Andrew Koppelman, "Why Discrimination against Lesbians and Gay Men Is Sex Discrimination," 69 New York University L. Rev. 197 (1994). See also Eric Heinze, Sexual Orientation: A Human Right (1995).

70. Cf. Margaret A. Farley, RSM, "An Ethic for Same-Sex Relations," in Robert Nugent, ed., A Challenge to Love: Gay and Lesbian Catholics in the Church 93, 105 (1983):

> My answer [to the question of what norms should govern same-sex relations and activities] has been: the norms of justice—the norms which govern all human relationships and those which are particular to the intimacy of sexual relations. Most generally, the norms are respect for persons through respect for autonomy and rationality; respect for relationality through requirements of mutuality, equality, commitment, and fruitfulness. More specifically one might say things like: sex between two persons of the same sex (just as two persons of the opposite sex) should not be used in a way that exploits, objectifies, or dominates; homosexual (like heterosexual) rape, violence, or any harmful use of power against unwilling victims (or those incapacitated by reason of age, etc.) is never justified; freedom, integrity, privacy are values to be affirmed in every homosexual (as heterosexual) relationship; all in all,

> individuals are not to be harmed, and the common good is to be promoted. The Christian community will want and need to add those norms of faithfulness, forgiveness, of patience and hope, which are essential for any relationships between persons within the Church.

71. By "lifelong" sexual unions, I mean sexual unions, whether heterosexual or homosexual, in which the partners hope and intend that their relationship will be lifelong, and in which they struggle with all the resources at their command to bring that hope and intention to fulfillment.

72. Mohr, n. 69, at 238.

73. For a statement of the Catholic Church's official position on homosexual sexual conduct, see Catechism of the Catholic Church 566 (par. 2357) (1994); see also id. (pars. 2358–59). See also Alan Cowell, "Pope Calls Gay Marriage Threat to Family," New York Times, Feb. 23, 1994, at A5. The "Ramsey Colloquium" agrees with the Catholic Church's official position. See "The Homosexual Movement: A Response by the Ramsey Colloquium," First Things, March 1994, at 15. For a powerful "counter-response", see the letter from various members of the National Association of College and University Chaplains, First Things, September 1994, at 2. For a discussion of the Church's official position, see Richard P. McBrien, Catholicism 993–97 (rev. ed., 1994).

For a sampling of influential critiques by Catholic theologians and philosophers of the position the Church espouses, see Farley, n. 70; Christine E. Gudorf, Body, Sex, & Pleasure: Reconstructing Christian Sexual Ethics (1994); Patricia Beattie Jung & Ralph F. Smith, Heterosexism: An Ethical Challenge (1993); Daniel Maguire, "The Morality of Homosexual Marriage," in Nugent, n. , at 118; Richard A. McCormick, SJ, Reflections on Moral Dilemmas Since Vatican II, ch. 17 ("Homosexuality as a Moral and Pastoral Problem") (1989); Richard Westley, Morality And Its Beyond 169–98 & 222–28 (1984). (Farley, Gudorf, Jung, Maguire, and McCormick are all Catholic moral theologians or ethicists; Jung's co-author, Smith, is a Lutheran pastor; Westley is a Catholic moral philosopher.) For an excellent discussion, see Jeffrey S. Siker, "Homosexual Christians, the Bible, and Gentile Inclusion: Confessions of a Repenting Heterosexist," in Jeffrey S. Siker, ed., Homosexuality in the Church: Both Sides of the Debate 178 (1994). (Siker is a Christian ethicist and an ordained member of the Presbyterian Church (USA).) See also David S. Toolan, "In Defense of Gay Politics: Confessions of a Pastoralist," America, Sept. 23, 1995, at 18. (Toolan, a Jesuit priest, is an associate editor of *America*, the Jesuit weekly.)

74. Andrew M. Greeley, "A Sea of Paradoxes: Two Surveys of Priests," America, July 16, 1994, at 6, 8.

75. Cf. Eric Zorn, "Citing a Wrong to Block a Right," Chicago Tribune, April 21, 1994, section 2, at 1:

> [T]he favorite biblical passage of those who rail against homosexuality [is] Chapter 18, Verse 22 of Leviticus: "You shall not lie with a male, as with a woman; it is an abomination."

> Suffice it to say that this particular book—with its obsession with animal sacrifice, expressions of disgust at the uncleanliness of menstruating women, approval of the death penalty for blasphemers, acceptance of human slavery, endorsement of torture, and vilification of the disabled—is not otherwise considered a reliable legislative guide in contemporary society.
>
> The Bible's relevance in such debates is further clouded by [the way in which] one can find in it justifications for any number of practices most of us frown on, including cannibalism (Deuteronomy 28), incest (Genesis 19), genocide (Numbers 31), self-mutilation (Matthew 18), and the execution of Sabbath breakers (Exodus 31).

76. See Brenda You, "A Holy War Against Gays," Chicago Tribune, April 26, 1994, section 5, p. 1.

77. For a sensitive elaboration of the point, see Thomas F. O'Meara, OP, Fundamentalism: A Catholic Perspective (1981).

78. See, e.g., Gerald D. Coleman, SS, "The Vatican Statement on Homosexuality," 48 Theological Studies 727, 733 (1987); Victor Paul Furnish, "The Bible and Homosexuality: Reading the Texts in Conext," in Siker, Homosexuality in the Church, n. 73, at 18; Jung & Smith, n. 73, ch. 3 ("The Bible and Heterosexism"); McBrien, n. 73, at 993–97; McCormick, n. 73; Siker, "Homosexual Christians, the Bible, and Gentile Inclusion," n. 73. See also note [Robinson].

79. See Jung & Smith, Heterosexism, n. 73; Virginia Ramey Mollenkott, "Overcoming Hetrosexism—To Benefit Everyone," in Siker, Homosexuality in the Church, n. 73, at 145.

80. See nn. 72–74, 78.

81. See John M. Finnis, "Law, Morality, and 'Sexual Orientation'," 9 Notre Dame J. L., Ethics & Public Policy 11, 16 (1995).

82. For other critiques of Finnis's argument (critiques that were drafted at the same time my own critique was drafted), see Andrew Koppelman, "Homosexuality, Natural Law, and Morality," in Robert P. George & Andrew Koppelman, eds., Homosexuality and Natural Law (forthcoming); Paul J. Weithman, "Natural Law, Morality and Sexual Complementarity," in Martha C. Nussbaum & David Estlund, eds., Love and Nature (forthcoming, 1996). (My critique was first published, in a longer version, in 1995: Michael J. Perry, "The Morality of Homosexual Conduct: A Response to John Finnis," 9 Notre Dame J. L., Ethics & Public Policy 41 (1995).)

83. Finnis, n. 81, at 16.

84. Id. at 17. In an effort to show that his argument is not a narrowly Christian argument, much less a narrowly Catholic argument, much less a narrowly Thomist argument, Finnis examines at length the views of some of the classical Greek philosophers, including Socrates, Plato, and Aristotle, all of whom, according to Finnis, "reject[ed] all homosexual conduct". Id. at 25. "All three of the greatest Greek philosophers, Socrates, Plato and Aristotle, regarded homo-

sexual *conduct* as intrinsically shameful, immoral, and indeed depraved or depraving." Id. at 17; see id. at 16–25. (Finnis also claims that "Immanuel Kant . . . likewise rejected all homosexual conduct . . . ." Id. at 25.)

Much of Finnis's essay is an argument, principally with Martha Nussbaum, about what the classical philosophers did or did not say about the morality of homosexual sexual conduct. (See also Gerard V. Bradley, "In the Case of Martha Nussbaum," First Things, June/July 1994, at 11.) The reader interested in pursuing that aspect of Finnis's essay should consult Martha C. Nussbaum, "Platonic Love and Colorado Law: The Relevance of Ancient Greek Norms to Modern Sexual Controversies," 80 Virginia L. Rev. 1515 (1994) (McCorkle Lecture). Appendix 4 to Nussbaum's essay is co-authored by Nussbaum and Kenneth J. Dover. Appendix 4 begins: "Because we believe it is very important to counter erroneous accounts of ancient Greek homosexuality, and because Professor Finnis's citation of Dover as if he supports Finnis's position has made public clarification of Dover's position urgent, we jointly state our position below." Finnis, "Law, Morality, and 'Sexual Orientation'," at 1641.

In denying that "the judgment that homosexual conduct is morally wrong is inevitably a manifestation . . . of mere hostility to a hated minority," Finnis is responding to views like Richard Posner's. See Richard A. Posner, Sex and Reason 346 (1992) (emphasis added):

> [S]tatutes which criminalize homosexual behavior express *an irrational fear and loathing of a group that has been subjected to discrimination, much like that directed against the Jews, with whom indeed homosexuals—who, like Jews, are despised more for who they are than for what they do—were frequently bracketed in medieval persecutions*. The statutes thus have a quality of invidiousness missing from statutes prohibiting abortion or contraception. The position of the homosexual is difficult at best, even in a tolerant society, which our society is not quite; and it is made worse, though probably not much worse, by statutes that condemn the homosexual's characteristic methods of sexual expression as vile crimes . . . . There is a gratuitousness, an egregiousness, a cruelty, and a meanness about [such statutes].

85. See John M. Finnis, Natural Law and Natural Rights (1980).

86. My focus here is on what Finnis says about the morality of homosexual sexual conduct, not on what Finnis says other writers said about the morality of such conduct. See n. 84.

87. Finnis states that his argument "is an application of the theory of morality and natural law developed over the past thirty years by Germain Grisez and others. A fuller exposition can be found in the chapter on marriage, sexual acts, and family life, in the new second volume of Grisez's great work on moral theology." Finnis, "Law, Morality, and 'Sexual Orientation'," n. 81, at 25. Finnis then cites, in fn. 35, to: "2 Germain Grisez, The Way of the Lord Jesus, *Living a Christian Life* 555–574, 633–680 (1993)."

88. Finnis, "Law, Morality, and 'Sexual Orientation'," n. 81, at 11. By ho-

mosexual conduct, Finnis means "bodily acts, on the body of a person of the same sex, which are engaged in with a view to securing orgasmic sexual satisfaction for one or more of the parties." Id. at 17.

89. Id. at 16.

90. Id. at 14. Finnis begins his argument by stating "three fundamental theses" that are "[a]t the heart of the Platonic-Aristotelian and later ancient philosophical rejections of all homosexual conduct". The first two theses are, as we will see, the crucial ones.

> (1) The commitment of a man and a woman to each other in the sexual union of marriage is intrinsically good and reasonable, and is incompatible with sexual relations outside marriage. (2) Homosexual acts are radically and peculiarly non-marital, and for that reason intrinsically unreasonable and non-marital. (3) Furthermore, according to Plato, if not Aristotle, homosexual acts have a special similarity to solitary masturbation, and both types of radically non-marital act are manifestly unworthy of the human being and immoral.

Id. at 25. Finnis then announces that he "want[s] to offer an interpretation of the three theses which articulates them more clearly than was ever attempted by Plato or, so far as we can tell, by Aristotle." Id.

Note that thesis (1) consists of two claims. The first is that "the commitment of a man and a woman to each other in the sexual union of marriage is intrinsically good and reasonable". (I heartily agree.) The second claim is that "the commitment of a man and a woman to each other in the sexual union of marriage . . . is incompatible with sexual relations outside marriage"—i.e., outside *such* a marriage, a *heterosexual* marriage. The phrase "is incompatible with sexual relations outside [heterosexual] marriage" does not mean "is incompatible with either the man or the woman having sexual relations outside their marriage". (Thus understood, the second claim would be quite irrelevant to the question of the morality of homosexual sexual conduct between two persons neither of whom is married to another and who might even consider themselves bound to one another in a lifelong, monogamous relationship of faithful love.) It means, rather, "is incompatible with *anyone* having sexual relations outside the context of a heterosexual marriage". Thus understood, the second claim is directed not specifically against homosexual sexual conduct, but against *any* sexual conduct, heterosexual as well as homosexual, outside the context of a heterosexual marriage. Thesis (2) presupposes thesis (1) as a major premise. If "the [intrinsically good and reasonable] commitment of a man and a woman to each other in the sexual union of marriage" were incompatible with anyone having sexual relations outside the context of a heterosexual marriage, then "[h]omosexual acts", which are an instance of sexual relations outside the context of a heterosexual marriage, would be incompatible with "the [intrinsically good and reasonable] commitment of a man and woman to each other in the sexual union of marriage".

91. Id. at 26.

92. Id. at 27.

93. I am ignoring, for present purposes, this complication: Many gay and lesbian couples do raise families; sometimes they do so because they have freely sought and happily embraced the opportunity to do so. Indeed, some gay and lesbian couples raise children who were begotten to be raised by them. On the subject of gay and lesbian parenting, see, e.g., Kimberly Lenz, "We Are Family: Gay and Lesbian Parents Face Challenges and Harbor Familiar Hopes—Growth as Loving, Committed, Open Families," ChicagoParent, May 1994, at 21.

94. Finnis, "Law, Morality, and 'Sexual Orientation'," n. 81, at 28.

95. Id. at 28–29. Finnis says, in the footnote (fn. 47) attached to this passage: "For the whole argument, see [Germain] Grisez, [2 The Way of the Lord Jesus, *Living a Christian Life* (1993),] at 634–39, 648–54, 662–64." See n. 87.

96. See Rich Heffern, "We Are Makers of Love: Interview with Richard Westley," Praying, May-June 1994, at 28.

97. Finnis, "Law, Morality, and 'Sexual Orientation'," n. 81, at 31 n. 50.

98. In choosing to satisfy one's sexual appetite, whether that appetite be heterosexual or homosexual—even in choosing to satisfy it in a "deliberately contracepted" way—one is not catering to or indulging a pathological appetite the satisfaction of which is always or even usually antithetical to one's authentic flourishing as a human being. Consider here Thomas Aquinas's teaching on homosexual activity (*in coitu masculorum*):

> Now with regard to pleasures of either of these two kinds, there are some which are unnatural, absolutely speaking, but may be called natural from a particular point of view (*sed connaturales secundum quid*). For is sometimes happens that one of the principles that is natural to the species as a whole has broken down in one of its individual members; the result can be that something which runs counter to the nature of the species as a rule, happens to be in harmony with nature for a particular individual (*fieri per accidens huic individuo*), as it becomes natural for a vessel of water which has been heated to give out heat. Thus something which is "against human nature," either as regards reason or as regards physical preservation, may happen to be in harmony with the natural needs of *this* man because in him nature is ailing. He may be ailing physically: either from some particular complaint, as fever-patients find sweet things bitter, and vice versa; or from some dispositional disorder, as some find pleasure in eating earth or coals. He may be ailing psychologically, as some men by habituation come to take pleasure in cannibalism, or in copulation with beasts or with their own sex (*in coitu bestiarum aut masculorum*), or in things not in accord with human nature.

Summa Theologica 1–2, 31–39, quoted in Coleman, n. 78, at 733. What is the contemporary Thomist to conclude if she comes to accept what was unknown to Thomas (as well as, of course, to the biblical authors), namely, that a homosexual sexual orientation, which is probably innate (as Finnis acknowledges; see n. 90 and accompanying text), is not the yield of one or another "psychological ailment" and, moreover, that acting in accord with one's homosexual sexual orientation is not necessarily antithetical to one's flourishing as a human being? (This is not to deny that homosexuals, too—especially homosexuals who have born the many terrible burdens of a culture that is pervasively homophobic—may ail psychologically.) Cf. McCormick, n. 73, at 300: "[O]fficial Catholic rejection of homosexual acts antedates by far knowledge of homosexuality as a not-chosen and most often irreversible orientation. That leads to the interesting and provocative question: does this knowledge have *no influence whatsoever* on the assessment of homosexual behavior (acts) at the objective level? One has to wonder if the distinction between orientation and acts, acknowledged now by all official documents and pastorals, has not remained abstract and unexamined in these documents with respect to its possible implications."

99. Why is conduct immoral simply because it involves one in treating one's body "as an instrument to be used in the service of one's consciously experiencing self"? Assume that from time to time I choose to eat a food—perhaps always the same food—that is utterly without nutritional value (and so does me no physical good) but that is otherwise harmless and satisfies my appetite for a particular taste or sensation. Assume, too, that I do not thereby fail to eat, or make it more likely that someday I will fail to eat, the nutritional foods I need. Have I thereby done something that "dis-integrates me precisely as an acting person"? Cf. Finnis, Natural Law and Natural Rights, n. 85, at 87 (discussing "play" as "[t]he third basic aspect of human well-being").

100. Finnis, "Law, Morality, and 'Sexual Orientation'," n. 81, at 28.

101. See n. 87.

102. See n. 87.

103. Grisez, n. 87, at 636 (emphasis added).

104. See McBrien, n. 73, at 982–92. Cf. Heffern, "We Are Makers of Love," n. 96, at 32 (quoting Richard Westley): "Love making, good mindful sexuality is a spiritual art. It is the most difficult of the spiritual arts. Making love is a human enterprise, not a blind biological urge. What makes it difficult is that it *can* be just a blind biological urge. It takes discipline, commitment and hard work to wring the good growth out of it that is possible. The spiritual art is to take this biological thing and transform it into a growth toward love, to bring forth the great potential."

105. In an effort to achieve "the most charitable reconstruction of Finnis's argument", Paul Weithman states: "Finnis's argument against homosexuality . . . depends on [the] claim . . . that human beings must never choose to act against the good realized by the 'biological (hence personal) units' constituted by voluntary, uncontracepted, heterosexual union. . . . [G]oods do seem worthy of respect and Finnis's claim that it is unreasonable to choose against them is plausible." (Weithman then adds that he believes that Finnis is wrong to conclude "that homosexual unions cannot actualize common goods.") Paul J. Weithman, "A Propos of Professor Perry: A Plea for Philosophy in Sexual Ethics," 9 Notre Dame J. L., Ethics & Public Policy 75, 79 (1995). No doubt, a (true) good *is* worthy of respect, and it is at least presumptively unreasonable to choose against a good. But goods in abstracto are not goods at all; they are intellectual construc-

tions. Goods, if they are goods at all, are concrete. And a concrete state of affairs, like "my conceiving a child" or "my getting pregnant", might be good for one or another person *but not good—even, perhaps, very bad—for one or another other person.* It is *not* unreasonable, even presumptively, for me to choose against a concrete state of affairs that is not good for me. (Of course, if my choosing against a concrete state of affairs that is not good for me involves my choosing against a concrete state of affairs that *is* good for someone else—or even choosing for a concrete state of affairs that is bad for someone else—the matter is more complicated. Choosing to have an abortion, for example, is much more problematic than choosing to engage in contracepted sex.)

106. Frank Rich, "Beyond the Birdcage," New York Times, March 13, 1996, at A15.

107. Finnis, "Law, Morality, and 'Sexual Orientation'," n. 81, at 29–30.

108. Thomas H. Stahel, SJ, "Transcending Biology," Commonweal, Mar. 11, 1994, at 2. Then, responding to the claim that "[f]or incarnational religion the intrinsically nonprocreative nature of homosexual acts is a metaphysical dead-end", Father Stahel continues: "Well, then, what about the intrinsically non-procreative nature of my vowed chastity, my clerical celibacy? And as long as we are on the subject of the Incarnation, what are we to say of Jesus Christ?—who was born of a woman, all right, but not by a heterosexually procreative act?" Id.

109. Martha Nussbaum has suggested, in correspondence, that "one might add that Finnis's contrast between judgment and emotion would not win the approval of *any* ancient thinker; besides, it's bad philosophy." For a book-length elaboration, see Martha C. Nussbaum, The Therapy of Desire (1994).

110. Finnis, "Law, Morality, and 'Sexual Orientation'," n. 81, at 31.

111. See, e.g., James Brooke, "With Church Preaching in Vain, Brazilians Embrace Birth Control," New York Times, Sept. 2, 1994, at A1: "In a country where Catholics account for 75 percent of the nation's 154 million people, every relevant statistic shows that most people ignore the Catholic Church's teachings on family planning methods. In a survey of 2,076 Brazilian adults in June, 88 percent of respondents said they 'don't follow' church teachings on birth control . . . Among women from 25 to 44, the 'don't follow' group expanded to 90 percent."

112. See n. 74.

113. Mahoney, n. 42, at 171.

114. Farley, n. 70, at 99–100. Paul Weithman writes: "If Perry recognizes the possibility that large numbers of people were in the grip of a sexually conservative illusion (as Russell and Lawrence alleged), then he must recognize the possibility that large numbers of people are in the grip of a sexually liberal one (as Finnis alleges)." Weithman, n. 105, at 84. I do recognize the latter possibility. Indeed, I more than recognize it: I believe that large numbers of people in our society and elsewhere *are* in the grip of one or another "sexually liberal" illusion, at least for some period of their lives—and some of them for their entire lives. However, what particular illusion is it that deliberately contracepting married couples—or lifelong, monogamous, faithful, loving homosexual cou-

ples—might be in the grip of? I just don't discern the possible—i.e., the *realistically* possible—illusion. (Does Weithman?) The issue doesn't seem to me a serious one. See Andrew Sullivan, "Alone Again, Naturally: The Catholic Church and the Homosexual," New Republic, Nov. 28, 1994, at 47, 55:

> [T]o dismiss the possibility of a loving union for homosexuals at all—to banish from the minds and hearts of countless gay men and women the idea that they, too, can find solace and love in one another—is to create the conditions for a human etiolation that no Christian community can contemplate without remorse. What finally convinced me of the wrongness of the Church's teachings was not that they were intellectually so confused, but that in the circumstances of my own life—and of the lives I discovered around me—they seemed so destructive of the possibilities of human love and self-realization. By crippling the potential for connection and growth, the Church's teachings created a dynamic that in practice led not to virtue but to pathology; by requiring the first lie in a human life, which would lead to an entire battery of others, they contorted human beings into caricatures of solitary eccentricity, frustrated bitterness, incapacitating anxiety—and helped to perpetuate all the human wickedness and cruelty and insensitivity that such lives inevitably carry in their wake. These doctrines could not in practice do what they wanted to do: they could not both affirm human dignity and deny human love.

115. Finnis, "Law, Morality, and 'Sexual Orientation'," n. 81, at 31.
116. See n. 93.
117. See n. 96 and accompanying text.
118. Finnis, "Law, Morality, and 'Sexual Orientation'," n. 81, at 32.
119. Id. at 29.
120. See McBrien, n. 73, at 983.
121. Joan Sexton, "Learning from Gays," Commonweal, June 17, 1994, at 28. Sexton also wrote, in the opening of her letter, that "the list of threats to modern Christian marriage is so long that gay marriage should rank about twenty-fourth, even for those who take it seriously."
122. Grisez and Finnis's particular argument against homosexual sexual conduct, connected as it is to their argument against "deliberately contracepted" sexual activity, is not idiosyncratic. That connection or linkage represents an official position of the Catholic Church, which teaches that a fundamental moral problem with all homosexual sexual conduct is its nonprocreative character. Cf. Michael J. Farrell, "Feisty New Ireland Leaves the Church Panting to Keep Up," National Catholic Rptr., July 29, 1994, at 7, 8 (quoting Archbishop Desmond Connell of Dublin, Ireland): "If, under the influence of the contraceptive culture, society accepts a view of marriage that releases the married couple from all commitment to procreation, it opens the way to the final debasement of marriage, the recognition of so-called homosexual marriages." On contemporary talk among Catholics about "the contraceptive culture", see Helen Fitzgerald,

"Needed: More, Not Less Talk About Sexuality," Commonweal, Nov. 4, 1994, at 42. Cf. Lance Morrow, "A Convert's Confession," Time, Oct. 3, 1994, at 88: "American Catholics—and millions elsewhere—understand that the church is simply out to lunch on the subject of birth control."

123. Rosemary R. Ruether, "The Personalization of Sexuality," in Eugene C. Bianchi & Rosemary R. Ruether, eds., From Machismo to Mutuality: Essays on Sexism and Woman-Man Liberation 70, 83 (1976). See also Nussbaum, n. 84, at 1530 (commenting on "the secular argument of Roger Scruton").

Weithman writes: "Why not think that very different moral considerations are relevant to the questions of whether to engage in orgasmic homosexual activity and whether to engage in heterosexual sex in a way designed to prevent conception?" Weithman, n. 105, at 88. I think I understand the point Weithman is trying to make, but I would put it differently. Imagine three couples: (1) a heterosexual couple who are able to conceive and bear a child in the normal way and who know that they have that ability; (2) a heterosexual couple who are unable to conceive a child and who know that they have that disability; and (3) a homosexual couple. Couple (1) faces a question that simply does not arise for couple (2) or couple (3): Whether to engage in sex in a way designed to prevent conception? Obviously the considerations that bear on this question are different from the considerations that bear on the question whether to engage in sex at all, which is a question that can and might arise for all three couples. (Does Weithman mean to say more than this? In particular, does Weithman mean to allow for the possibility of a plausible moral perspective from which the following claim is defensible: "Even if orgasmic homosexual sexual conduct is not always morally wrong, deliberately contracepted heterosexual conduct *is* always morally wrong.")

124. Anthony of the Desert, quoted in Thomas Merton, ed., The Wisdom of the Desert 62 (1960). St. Anthony was a fourth-century Christian monk.

125. See preface, n. 10.

126. See generally Hunter, n. 38.

127. David Hollenbach, SJ, "Contexts of the Political Role of Religion: Civil Society and Culture," 30 San Diego L. Rev. 877, 894 (1993).

128. Id. at 894–95.

129. David M. Smolin, "The Enforcement of Natural Law by the State: A Response to Professor Calhoun," 16 U. Dayton L. Rev. 318, 391–92 (1991).

130. Mark A. Noll, The Scandal of the Evangelical Mind (1994).

131. Id. at 207–08.

132. See n. 124.

133. See n. 48.

134. See n. 47 and accompanying text.

135. John H. Robinson, "Church, State, and Sex," 9 Notre Dame J. L., Ethics & Public Policy 1, 5 (1995).

136. This is not even to say that a believer's principal reliance should be on the secular argument.

137. Michael J. Perry, Morality, Politics, and Law 181–82 (1988). See also Perry, Love and Power, n. 49, at 4.

138. Cf. Perry, Morality, Politics, and Law, n. 137, at 183: "If one can participate in politics and law—if one can use or resist power—only as a partisan of particular moral/religious convictions about the human, and if politics is and must be in part about the credibility of such convictions, then we who want to participate, whether as theorists or activists or both, must examine our own convictions self-critically. We must be willing to let our convictions be tested in ecumenical dialogue with others who do not share them. We must let ourselves be tested, in ecumenical dialogue, by convictions we do not share. We must, in short, resist the temptations of infallibilism."

139. Mahoney, n. 42, at 327 (emphasis added).

140. Noonan, n. 48, at 676–77.

141. For relevant discussions, see Stephen L. Carter, "The Religiously Devout Judge," 64 Notre Dame L. Rev. 932 (1989); Sanford Levinson, "The Confrontation of Religious Faith and Civil Religion: Catholics Becoming Justices," 39 DePaul L. Rev. 1047 (1990); Lawrence B. Solum, "Faith and Justice," 39 DePaul U. L. Rev. 1083 (1990); Howard J. Vogel, "The Judicial Oath and the American Creed: Comments on Sanford Levinson's *The Confrontation of Religious Faith and Civil Religion: Catholics Becoming Justices*," 39 DePaul L. Rev. 1107 (1990); Thomas L. Shaffer, "On Checking the Artifacts of Canaan: A Comment on Levinson's '*Confrontation*'," 39 DePaul L. Rev. 1133 (1990); James L. Buckley, "The Catholic Public Servant," First Things, February 1992, at 18; Scott C. Idleman, "The Role of Religious Values in Judicial Decision Making," 68 Indiana L. J. 433 (1993).

142. For a classic discussion of the point, see H.L.A. Hart, The Concept of Law 124–36. (2d ed. 1994) (on "the open texture of the law").

143. See Lawrence B. Solum, "On the Indeterminacy Crisis: Critiquing Critical Dogma," 54 U. Chicago L. Rev. 462 (1987).

144. On the judicial "specification" of underdeterminate constititional materials, see Perry, The Constitution in the Courts, n. 11, chs. 5 & 6.

145. John Rawls, Political Liberalism 216 (1993).

146. Id.

147. Id. at 236.

148. See ch. 2, section III (commenting on the problem that such underdeterminacy poses for Rawls's approach).

149. Let me emphasize, to avoid misunderstanding, that my position is *not* that courts may rely on nonlegal norms to "trump" or override the relevant, legally authoritative norms. The question concerns the permissible bases, not for overriding the relevant, legally authoritative norms, but for shaping them in contexts in which they happen to be underdeterminate. See Perry, The Constitution in the Courts, n. 11, at 98–99:

> The heart of the judge's responsibility, of course, is to decide by relying on "legal" premises, if there are any relevant legal premises: premises authoritative for her *qua* judge. Assuming that the relevant legal premises do not conclude the question, it seems fitting for a judge to

decide by relying on premises that, although not authoritative for her *qua* judge, are nonetheless the object of widespread consensus in American society—even, perhaps, part of the society's "common sense"—unless the consensus/common sense is, in her view, either contrary to legal premises or mistaken; conversely, it seems problematic for her to decide by relying on premises widely rejected in American society. According to Justice Brennan, "[E]ven high court judges are constrained in issuing rulings[,] . . . not just by precedent and the texts they are interpreting, but also, *on any attractive political and jurisprudential theory*, by a decent regard for public opinion. . . ."

Assuming, however, that legal premises and/or consensual/commonsensical premises, even if they rule out some answers to the "how best to achieve the value question", do not yield a single answer, presumably she should decide by relying on premises she accepts, premises authoritative for her *qua* the particular person she is—unless, of course, an axiomatic (for the political-legal culture) norm about judicial role requires her to forsake reliance on one or more premises she accepts (or unless one or more premises she accepts is widely rejected in American society). What sense would it make to suggest that when legal premises and consensual/commonsensical premises do not together yield an answer, a judge should rley on premises she rejects, premises not authoritative for her *qua* the particular person she is? (To say that an axiomatic norm about judicial role may require her to rely on one or more premises she rejects is just to say that one or more premises she rejects may be authoritative for her *qua* judge—that, in other words, one or more such premises might be legal premises.) As U. S. Circuit Judge James L. Buckley has written, in an essay titled "The Catholic Public Servant": "When faced with ambiguities, or with problems that fall within the interstices that inevitably exist within and between laws, a judge is necessarily called upon to exercise a large measure of discretion. In doing so, he will inevitably bring to that task everything that he is—the books he has read; his experience as spouse, parent, and public official; his understanding of the nature of man and the responsibilities of citizenship; his sense of justice; even his sense of humor. A judge is not a machine, and the judicial function cannot be displaced by a formula or measured by an equation." To say, as I did a moment ago, that a judge should not rely on premises widely rejected in American society is not to say that she should never rely on premises not widely accepted in American society; it is not to say that she should "never be the first person to bring a new value, a new political or ethical insight, into the law." As Justice Brennan has said, "High court judges interpreting a bill of rights may at times lead public opinion". (Justice Brennan quickly added, however, that "in a democratic society they cannot do so often, or by very much.")

See also Jeremy Waldron, "Religious Contributions in Public Deliberation," 30 San Diego L. Rev. 817, 833 (1993):

> [A] point will come in judicial decision making when the judge must simply make a moral judgment of his own, in his own voice, the best way he knows how. He does not *eo ipso* become a moral philosopher, unless we extend the latter category to cover anyone who tries to think hard about moral problems. It makes more sense to adjust the categories from the other direction: moral and political philosophers are simply doing systematically, professionally, and at leisure what judges and other officials must do under a deadline every day. Neither activity—neither that of the philosopher nor that of the judge—differs, in essence, from the thinking and decision making of the ordinary person addressing matters of civic importance. For all three, a time comes when he must think about and enter a value judgment in his own voice. Unless he does that, he will not be able to complete the task assigned to him, whether it is teaching, interpreting legal sources, or choosing for whom to vote. . . . Once we acknowledge the unavoidable place of moral judgment in the activity of the judge, we must recognize that he is *pro tanto* in the same game as the ordinary citizen. So, anything appropriate for the citizen to take into account in exercising his political power is appropriate for the judge and for other officials to take into account, to the extent that they face moral choices in the exercise of theirs.

150. Indeed, legislators and other public officials, including judges, should not rely on a religious premise in making a choice unless, in their view, a persuasive secular premise supports the choice. See ch. 1, n. 97.

151. See Greenawalt, Private Consciences and Public Reasons, n. 9, at 149–50.

152. See id. at 150.

153. Id.

154. Joseph Raz, "The Politics of the Rule of Law," 3 Ratio Juris 331, 331 (1990). With respect to legislation, the rule of law requires "that new laws should be publicly promulgated, reasonably clear, and prospective". Id. With respect to administration, it requires that the law be enforced against everyone (everyone, that is, to whom the law applies, everyone the law restrains or constrains), even the most powerful members of society, including the highest government officials, and that the equal protection of the law be extended to everyone, even the most marginal members of society, including the poor and the unpopular.

INDEX